TECHNICAL
INFORMATION
SOURCES

TECHNICAL INFORMATION SOURCES

A GUIDE TO PATENT SPECIFICATIONS, STANDARDS AND
TECHNICAL REPORTS LITERATURE

SECOND EDITION BY
BERNARD HOUGHTON MA FLA
SENIOR LECTURER IN INFORMATION WORK
DEPARTMENT OF LIBRARY AND INFORMATION STUDIES
LIVERPOOL POLYTECHNIC

LINNET BOOKS & CLIVE BINGLEY

FIRST PUBLISHED 1967 THIS REVISED EDITION FIRST
PUBLISHED 1972 BY CLIVE BINGLEY LTD
AND SIMULTANEOUSLY PUBLISHED IN THE USA BY
LINNET BOOKS, AN IMPRINT OF SHOE STRING PRESS INC
995 SHERMAN AVENUE, HAMDEN, CONNECTICUT 06514
PRINTED IN GREAT BRITAIN
COPYRIGHT © BERNARD HOUGHTON 1972
ALL RIGHTS RESERVED
0-208-01074-2

cop.2

Tec

CONTENTS

5,50

1*

INTRODUCTION

Although there are articles which are concerned with the handling of patents and technical reports these are scattered through a large number of documentation and technical periodicals. No suitable text exists which explains the nature and evolution of patents, standards and technical report literature to those who have never before encountered these documents but who will be confronted by them when they take up professional posts in technical libraries, information services and industry. I have attempted to fill this gap in the literature.

The work will also be useful to students of librarianship and information work who are preparing for examinations on degree courses or under the Library Association syllabus. It is essential that students fully appreciate the nature of these basic sources of information before approaching such subjects as abstracting and information retrieval. The book should also prove valuable to those librarians or academics in university and college libraries who are responsible for organising courses on the structure and use of scientific literature.

Since the first edition of this work was published in 1967 there has been much activity in the field of patents and standards. Substantial changes have either taken place or have been planned in the patent laws of several countries. International activity has resulted in the ratification of the Patent Cooperation Treaty, and real progress has been made towards the European patent. International cooperation in the standards sector has resulted in the establishment of organisations and committees charged with the tasks of harmonising both standardisation and licensing programmes. These developments represent the major expansions in this revised edition.

I would like to acknowledge my gratitude for the assistance I received in gathering material for the book from the staffs of the

7

National Reference Library for Science and Invention (particularly Mr D R Jamieson FLA), British Standards Institution, The National Lending Library for Science and Technology and Liverpool Technical Library and LADSIRLAC. Special thanks are also due to my colleagues, Mr Frank Gibbons MA FLA for reading my original manuscript and making some helpful suggestions, and Mr Glyn Rowland FLA for his constructive comments on the revised text. Finally I wish to acknowledge the help received from Mr G D McCall and Mr P M Gore, Chartered Patent Agents, on points of British and foreign patent procedure.

Gateacre, Liverpool   BERNARD HOUGHTON
October 1971

Chapter one: *The nature and evolution of patents and the British patent system*

*P*atent specifications are a basic source of information; in the United Kingdom alone an average of eight hundred specifications is published each week. Each specification is a detailed description of a new device or method of production followed by certain claims made on the part of the inventor. When the patent is eventually granted, the individual inventor is presented with ' letters patent ' which confer on him the right to restrain other people from using his invention without his consent. The letters patent is a contract between the state and the inventor, but the protection it affords is only granted for a limited period, sixteen years in the United Kingdom. In other countries this period varies from twelve to twenty years.

The purpose of the patent system is to liberate an inventor's ideas so that they can be used for the benefit of the community, but as a reward for his ingenuity the inventor is given the opportunity of being the first individual to put the invention to work and thus to reap the first profits. The system of ' compulsory licences ' which obtains in the United Kingdom underlines the fact that patents exist for the benefit of the community rather than for the enrichment of the inventor. If an inventor does not make reasonable attempts to put his patent to work within three years of its being granted any person is entitled to ask the Comptroller of Patents for a ' compulsory licence ' to work the invention.

The patent system is not of recent origin; its antecedents can be traced to the monopolies which were formerly the prerogatives of ruling monarchs. In 1236 Henry III of England granted to a citizen of Bordeaux the right to manufacture cloth of various colours for a term of fifteen years, after which time the monopoly was to lapse. Although other instances of the granting of monopolies were not uncommon in Western European countries in the thirteenth and fourteenth centuries, Venice can claim to be the first state to develop a regular patent system. The first Venetian patent grant which has been documented was made on 20 February 1416, and before 1550 almost one hundred monopolies varying from twenty to sixty years had been issued.[1]

Today patent systems are operating in most countries of the world. A complete list of over two hundred existing systems which is kept up to date by looseleaf supplements is W W White & B G Ravenscroft *Patents throughout the world* (Trade Activities Inc, 1959).

THE BRITISH SYSTEM

In Great Britain many of the earliest monopolies were granted to foreign tradesmen to encourage them to settle and to pursue their trades in this country. Although these grants were not patents they do illustrate that a system of government promotion of industrial development and protection of the rights of the individual existed from early times. The monopoly was introduced to encourage new trades and introduce new products but in practice monopolies were granted for the most commonplace articles; the system was abused and often a tradesman could not follow his chosen trade without infringing someone's monopoly. In 1598 Queen Elizabeth I enabled Edward Darcy, a member of her court, to make a considerable amount of money by granting him a monopoly of producing and selling playing cards, an old established commodity, for twenty one years. The system fell into further disrepute when the Stuarts used the monopoly as an instrument of political favour. These unfortunate grants led to a succession of conflicts between the people and the Crown and the need for reform was soon clear. The Statute of Monopolies of 1624 placed a limitation on the powers of the Crown and laid down certain conditions under which patents might be granted. Section 6 of the statute declared illegal ' all monopolies for the sole buying, selling, making, working or using of anything ', but the declaration did not extend to ' letters patent and grants of privilege for the term of fourteen years or under hereafter to be made of the sole working or making of any manner of new manufacture within this realm to the true and first inventor and inventors of such manufactures which others at the time of making such letters patents and grants shall not use so as also they be not contrary to the law nor mischievous to the State by raising prices of commodities at home or hurt of trade or generally inconvenient'. No further major legislation affected the law of patents until the Amendment Act of 1852 established the Patent Office as the central point for all matters dealing with patents in the British Isles. All British patent law and procedure is now governed by the Patents Act of 1949.

THE APPLICATION
An application for a British patent must be made by the inventor himself or by his assignee. The inventor is the person who originated the invention or who was responsible for its introduction into the British Isles. Any individual or company may apply for a patent as the assignee of the inventor and one or more persons may join as co-applicants with the inventor or his assignee. Any patent rights relating to discoveries made by a research worker or any other employee who is engaged by an organisation on research or in an inventive capacity will legally belong to the employer. A firm which does claim such patent rights is not legally bound to pay a remuneration to the research worker whose invention it utilizes, although many firms do operate bonus schemes as incentives to their research staff. If, however, the research worker invents a device or process during his period of employment which has no relation to the firm's scope of activity the firm would have no legal claim on this invention. Any rights to inventions produced by a machinist or operative while he is employed by a firm will belong to the individual and not the firm, since he is not employed in an inventive capacity. Most organisations now insist that all research staff complete a contract covering the ownership of any patent rights relating to their inventions.

Application for a patent must be made on a standard form which is obtainable from the Patent Office. A revised edition of the Office's pamphlet *Applying for a patent*, a non-specialist survey of British procedure was published in June 1971. The application must be accompanied by a specification, a description of the invention, which can be either provisional or complete. The British Patent Office is currently receiving 60,000 applications per year and two-thirds of this number are from overseas. When the application has been filed at the Patent Office it is given an application number and the date of filing is recorded. The application date is also known as the 'priority' date and if two individuals apply separately for patents for a similar invention, the patent will be granted to the individual filing the application first in time. Anytime after the application has been filed at the Patent Office the subject of the application may be used or published without prejudicing the applicant's chances of obtaining letters patent. However, premature disclosure should be avoided as the application may not be successful and thus any information disclosed could be used by competitors without redress.

INTERNATIONAL CONVENTION

The United Kingdom is a signatory to the *International convention for the protection of industrial property* which was originally ratified in Paris in 1883 and which has been revised several times since that date. A basic principle of the convention, which has been signed by more than sixty nations, is that a foreigner shall have the same patent rights as a national in any country signing the convention even if his own country has not signed the convention. The most important provision of the agreement is that a person who applies in a convention country for a patent enjoys in other convention countries the right to priority for a period of twelve months from the date of application in the original convention country. The Convention then ensures that an inventor who files an application in one country can publish information on the invention without the risk of his rights being pirated abroad provided that he files within the year a patent application in each of the countries in which he seeks protection.

A provisional specification, which will ensure an early priority date, can be made before finer points of detail have been worked out. The complete specification must be filed within twelve months, or fifteen months on payment of a fee, of the first application or that application will lapse. The complete specification must indicate the best method of putting the invention into practice; failure to comply with this condition could lead to the revocation of the patent in a subsequent action. Each specification must contain the specific claims which are being made by the inventor and must also be accompanied by drawings if these are necessary for an understanding of the invention.

Drafting of the specification, whether provisional or complete, is an exacting task calling for specialist knowledge. The specification must describe the invention precisely and the claims must neither omit anything that is essential to the invention nor add anything not actually invented by the applicant. As the inventor who prepares his own specification is almost certain to endanger his chances of obtaining a good patent, most inventors will employ a patent agent (patent attorney in the United States) whose function it will be to search for anticipatory documents, draft the specification, claims and drawings, and advise his client on replies to objections raised by the Patent Office; in short the patent agent will do everything within his power to ensure that his client is granted letters patent for his invention. Another important function of the agent is to ensure that his client does not infringe another's patent or if a patent is infringed to prove that the patent in question is not valid. The patent agent may also be engaged to notify a client of any new patents which may infringe or supersede his patents, to inform a client who may have several patents when their respective renewal fees are due, and to draft any necessary legal instruments such as deeds of assignment. The agent is not normally concerned with the financial exploitation of letters patent. In addition to patents, agents must be capable of acting on behalf of their clients on matters concerning other forms of industrial property such as trade marks and registered designs.

The patent agent is a technical solicitor who must have an exhaustive knowledge of patent law to enable him to foresee and counter any objections raised to the granting of the patent; considerable technical knowledge to enable him to appreciate the basic nature of the inventions of a wide range of clients and perhaps to suggest openings for their exploitation; and a fluent command of language to enable him to draft detailed legal documents. Only a chartered patent agent, who has passed a series of examinations to admit him to the register of the Chartered Institute of Patent Agents (CIPA) (Staple Inn Buildings, London WC1), or a solicitor who is on the register of the CIPA, is allowed to act on behalf of a client. The institute, which exists to 'promote improvements in the patent laws and to establish rules to be observed by patent agents in all matters relating to their professional practice ', publishes a copy of its register in February each year. The register gives for each agent his full name, date of registration, qualifications, and business address. At present there are some seven hundred and fifty chartered patent agents in Britain, two thirds of this number being freelance agents in private practice. Many large firms maintain patent departments and employ patent agents (known as ' tame agents ') to prosecute patent applications for all manufactures or processes which have been developed by the firm. A recent survey of two hundred and fifty American firms indicated that one third of this number employed patent attorneys as ' tame agents '.

The complete specification, when filed at the Patent Office, is examined to ensure that the invention has been fairly described and clearly illustrated, and that the title is sufficiently indicative of the subject content. A successful patent application must be :

1 Novel: a search is made by a Patent Office examiner who has a knowledge of the subject to which the specification relates. The search is usually limited to prior British patents of the last fifty years although the examiner may include any documents other than British patent specifications older than fifty years. The examiner will also write reports on specifications for submission to and subsequent discussion with the applicant or his agent. In cases of dispute he may be required to present the case against the application at the hearing. Over five hundred examiners are currently employed at the Patent Office.

2 A method of manufacture: something that can be manufactured or can assist in a manufacturing process.

3 Useful for the purpose for which it was intended.

4 Genuinely inventive: not obvious to someone skilled or experienced in the field to which the invention belongs. Under British patent law there is no obligation for the patentee to provide theoretical explanations as to why his invention works. The inventor does not need to describe his invention in more detail than is necessary for a person skilled in the art to make it work.

The successful application must not be :

5 Contrary to law or morality.

6 A recipe for a food or medicine simply consisting of a mixture of known substances and possessing only the sum of the known properties of the ingredients.

7 Contrary to natural laws, *eg* a perpetual motion machine.

If any of the above points are cited as objections by the examiner the applicant may be required to amend his specification.

OPPOSITION TO A PATENT GRANT

For three months after the completed specification is published any person who is financially interested in the invention (*eg* an industrialist who is about to manufacture a similar product), may oppose the granting of the patent under Section 14 of the 1949 Act. The most common grounds for opposition proceedings are:

1 Obviousness: the invention was obvious to anyone with a reasonable knowledge of the 'prior art'.

2 Prior publication, where the opposing party will produce a document of any date, with the exception of British patent specifications older than fifty years, to which the public has had access during the last fifty years which describes an invention identical to the subject of the patent application. The document must describe the invention completely, a composite picture of the invention built from several sources is not admissible as evidence of prior publication. The date-received stamp used on all documents added to the stock of the Patent Office Library is accepted as evidence that the publication was available to the public on a certain date.

3 Prior claim: the application is overlapped by the claims of a complete specification published on or after the date of filing the applicant's complete specification.

4 Prior use: the invention has been used publicly in the United Kingdom before the date of application for a patent. There are certain exceptions to the conditions of prior use, one being use by the applicant himself within one year of the date of application.

5 Obtaining: the applicant has obtained the invention from some person in the British Isles who is not mentioned in the application.

6 Essentially new matter has been introduced into the complete specification which was not foreshadowed in the provisional specification.

7 The invention has been insufficiently described.

8 In the case of a Convention application the British application was filed more than twelve months after the first application abroad.

9 The subject of the application is not an invention under the meaning of the Patents Act.

Approximately two per cent of all British patent applications are opposed.

When three months have expired without opposition, or when any opposition has been overcome, the patent is sealed and the patentee is presented with his letters patent, a document under the seal of the Patent Office signed by the Comptroller of Patents and accompanied by a printed copy of the complete specification. It is usually possible for a British patent to be granted and sealed within two years from the filing of the complete specification.

REVOCATION

During the twelve months following the grant of letters patent an application may be made by an interested party for the revocation of a patent under section 33 of the 1949 Act on grounds identical to those for opposition under section 14. Applications for revocation are heard by the Comptroller of Patents. A final safety mechanism of the British patent system is the machinery for revocation by the courts. Such proceedings can only be heard before the High Court (Chancery Division) and are comparatively rare, perhaps a dozen cases a year, as the cost of patent proceedings before the High Court is amongst the most expensive of all forms of litigation. The basis for revocation at this stage is wider than under sections 14 or 33; additional grounds include inutility, the invention will not work, and false suggestion, *ie* the procedures detailed in the specification cannot be fulfilled.

NUMBERING AND FEES

British patent specifications were numbered consecutively from the year 1617 to 1851, *ie* numbers 1-14359. From 1852 to 1915 a new sequence was commenced annually. The present sequence commenced at 100,001 in 1916 and specifications are numbered consecutively as they are issued.

Renewal fees become payable at the end of the fourth year and must be paid annually thereafter, otherwise the patent will lapse. Fees are on an increasing scale ranging from £13 for the fifth year to £40 for the sixteenth year, the total cost of obtaining patent protection for the full period of sixteen years being £300. Only in rare cases will the life of a patent be extended beyond sixteen years, if for instance wartime circumstances have prevented the exploitation of the patent, or if the patentee can prove that he has not received adequate remuneration from a patent which has been of exceptional public benefit.

PATENTS OF ADDITION
Patents of addition may be granted for modifications or improvements to inventions for which patents have already been granted, or to inventions for which a patent application has already been made; but the application for patents of addition must be made in the name of the original patentee or applicant. A patent of addition is subject to renewal fees but it will only remain in force as long as its parent patent is valid.

The granting of a patent gives the patentee the sole right to produce or utilise his invention and to restrain others from doing so. The patentee may however assign his patent to another party or allow that party to utilise his patent under licence. Assignment transfers the ownership of the patent but licensing merely allows the second party to use the patentee's invention under agreed terms. The central government is in a privileged position and may authorise any of its departments to utilise a patent in the interests of the Crown, without requesting the patentee's consent. If this prerogative is claimed the patentee will, of course, be compensated.

THE BANKS COMMITTEE
In 1967 the President of the Board of Trade set up a committee to review the existing British patent system. The Banks Committee's terms of reference were ' to examine and report with recommendations upon the British patent system and British law, in the light of the increasing need for international collaboration in patent matters and in particular of the UK government's intention to ratify the recent Council of Europe's Convention of patent laws '.[2]

The report of 1970 recommended that major changes should be introduced into the British system. Some of the important recommendations of the Banks Committee report are summarised below.

1 A fixed maximum period of three and a half years should be established from the earliest priority date to simultaneous grant and publication.

2 The publication, within eighteen months of the first priority date claimed, of all complete specifications as filed.

3 The separation of the novelty search from examination for patentability.

4 The novelty search would be made within six months of the filing of the complete specification, and the result of the search would be published at the date of early publication.

5 A request for examination, accompanied by a substantial fee, could be filed after the publication of the search report, but this application could be delayed up to two and a half years after the priority date.

6 A third party should be able to file, at his own expense, a request for examination, but prosecution would remain a matter between the applicant and the Patent Office.

7 The pre-grant opposition procedures would be repealed, as it is claimed that this would expedite grant.

8 The scope of the British novelty search should be extended beyond prior British specification of the previous fifty years and should also include US specifications and other forms of publication.

9 Search reports issued by the International Patent Institute (IIB) should be considered as official by the British Patent Office.

10 The British examiners would be able to reject an application on the grounds of obviousness.

It is claimed that three main advantages would accrue from the implementation of the above proposals. A patent granted under the revised system would have a greater chance of validity as a result of the more rigorous search procedure. The information would be disclosed more rapidly because of the early publication arrangements. Finally the patent would be granted within a fixed period of time and grant could not be delayed by protracted opposition proceedings.

REFERENCES
[1] Mandich, G: *Journal of the Patent Office Society* 42 (6) June 1960 379.
[2] *The British patent system. Report of the committee to examine the patent system and patent laws.* HMSO 1970. Cmnd 4407.

Chapter two: Overseas patent systems

UNITED STATES OF AMERICA

The first US patents were issued under the Patent Act signed by George Washington in 1790, although before this date individual American colonies and states had issued their own patents, the first on record being granted in 1641 by the Massachusetts general court to Samuel Winston for a method of manufacturing salt. Thomas Jefferson, third President of the USA and a major figure in American patent history, was a member of the patent board which granted the first US patent to Samuel Hopkins of Vermont for 'making pearl ashes'. The United States Patent Office was instituted under a Commissioner of Patents in 1836. The US patent system developed rapidly, the annual output of specifications increasing from three in 1790 and thirty three in 1791 to forty five thousand in 1964. Some three million US patents have now been issued since the institution of the system. The US Patent Office now handles more patent applications (about 100,000) a year than any other country. The ever increasing volume of applications inevitably causes delays in the granting of patents. The average time for a patent application to be pending in the United States is now three years, and in 1965 a presidential committee was instituted to overhaul the patent system and to examine possible ways of reducing the pending period.

Procedure: United States patent procedure is governed by the law passed on 1 January 1953. In the United States it is possible to patent any new, useful and unobvious machine, method of manufacture, useful art, composition of matter or asexually produced plant. The application must be accompanied by a specification which adequately describes the invention and includes specific claims made by the inventor and drawings if the invention can be illustrated. It is not possible to file a provisional specification in the United States. The application must be made in the name of the true inventor or inventors; no application can be made in the name of an assignee and no invention can be imported from abroad. Once the application has been filed at the US Patent Office it cannot be modified by the introduction of any essentially new matter.

All patent applications are examined for novelty at the US Patent Office; the examination is exceptionally thorough and it is not unusual for a foreign inventor to file an application in the United States simply to assure himself that his invention has not been anticipated. The application is also checked to ensure that it has been adequately described and that it is not obvious to anyone with a reasonable knowledge of the art. The application will not be granted if :

1 The subject of the application was known or used by others in the United States or patented or described in a printed publication in the United States or a foreign country before the invention of the device by the applicant.

2 It is patented or described in a printed publication in the United States or a foreign country or in public use or sale in the United States more than one year prior to the date of the application for the patent in the United States.

3 The application has been abandoned by the inventor.

4 It has been patented or a patent has been applied for by the applicant or his assignee in a foreign country more than twelve months previously.

5 The applicant did not invent the subject matter of the application.

6 The application was preceded by a similar application in the United States; here an interference is declared and legal proceedings ensue to determine priority. The foreign applicant can only claim a foreign convention date as priority, but the native applicant can trace his dates back as far as records such as laboratory notebooks will allow.

7 The subject matter of the application would be obvious to anyone having an ordinary skill in the art to which the invention pertains.

When the examination has been completed at the Patent Office, any objections raised by the examiner are communicated to the applicant in an ' official action '. This action, which must be answered within six months or otherwise the application will lapse, will cite any relevant patents or other publications. The applicant's answer will be in the form of an ' amendment ' which will withdraw or alter any offending claim. If the application has been successfully negotiated, letters patent are granted to the applicant. There are no renewal fees to be paid and the patent will remain in force for seventeen years from its date of issue.

There is no provision for an opposition period between the acceptance of the specification and the granting of the letters patent, as in the British system. Any suggestions to introduce some form of opposition procedure into the American system in the past have been opposed by patent law organisations on such grounds as that these proceedings would cause unnecessary delays in the prosecution of patent applications. The American patentee is under no obligation to work his invention, as there is no compulsory licensing system in operation under United States patent law. The owner of a US patent may, of course, sell or transfer his patent rights to another party by completing the necessary legal documents.

The steadily increasing number of patent applications being received by the US Patent Office each year has caused great concern. The main problem has been the accumulation of applications awaiting examination, and the consequent increasing delays between the receipt of the application and the publication of the corresponding specification. In 1965 President Johnson appointed a committee to review the US patent system and this body reported in December 1966. The committee was less than satisfied with existing procedures and recommended changes which were subsequently written into a draft 1967 Patent Reform Act. This Act sought to a) raise the quality and reliability of US patents; b) reduce the time and expense of obtaining and protecting a patent; c) speed public disclosure of scientific and technical information.

The main recommendations of the Act were :

1 The introduction of a first-to-file basis for awarding US patents. It was claimed that it was more equitable to grant a patent to the inventor ' who first appreciated the worth of the invention and promptly acted to make the invention available to the public '. Under existing US practice priority can be established by ' reduction to practice ' or ' date of conception ' rather than by date of application. This often results in long and involved ' interference

proceedings' between opposing applicants and also encourages the large patent-conscious corporations to maintain elaborate records of their work, which gives them the advantage in interference proceedings over smaller organisations and individuals who are usually unaware of the intricacies of patent procedure. It was argued that a first-to-file basis would encourage earlier public disclosure and would, in addition, bring US practice into line with that of most other countries.

2 The inventor would be allowed to file an informal technical disclosure which would establish his priority and provide him with a period in which to refine his invention before submitting a complete specification.

3 All pending applications would be published within eighteen to twenty four months.

4 An applicant could request that his application be published immediately, but the search could be delayed for up to five years. If by this time neither the inventor nor a third party had requested a search the subject of the application would be freely available for public use. As in all systems embracing deferred examination this, it was intended, would relieve the pressure from the examining personnel in that inventions of little economic significance would never be examined.

5 The life of the patent would be twenty years from the date of the application.

The 1967 Reform Act aroused the hostility of the state patent law associations and the industrial patent community. In the face of their opposition a weakened reform Bill was introduced in 1969, which dropped the first-to-file clause, deferred examination, and the provision which would have made disclosure mandatory within eighteen months. The main features of this Bill, which is still under consideration, are the introduction of both opposition procedures and renewal fees. If the Bill becomes law, members of the public will be permitted to cite prior art, including evidence of public use, to the Patent Office to defeat the grant of a patent.

Although reforms of the existing system have been delayed, the US Patent Office has been able to introduce some changes, referred to as defensive publication, in its procedures to reduce its workload. An applicant can now inform the office, within a limited period after filing, that he no longer wishes to pursue his application. The office will then publish an abstract of the application and lay open the file to public inspection. The applicant will forego any rights to a monopoly, but the publication of his application will form a defence against subsequent applications by others seeking to deny the original inventor the right to utilise his invention.

CANADA

The Canadian system is an amalgam of American and British practice. The invention must not have been described in a publication printed anywhere more than two years before filing, or have been in public use or on sale in Canada or elsewhere for more than two years. The application is examined for novelty, but no formal opposition can be lodged. The Canadian Patent Office will acknowledge a protest against a specification, but will not give details of subsequent action taken. The patent will remain in force for seventeen years from the date of grant and no renewal fees are necessary to keep it in force. If the invention is not worked within three years of the grant, any person may apply for a compulsory licence, as under British law. Similar applications for a patent are decided by interference or ' conflict ' proceedings which are heard by the Commissioner of Patents.

EARLY PUBLICATION

Belgian, French and German patent specifications are important sources of technical information because of their early disclosure dates. These are made possible because the specifications are published before they have been examined for novelty. Under French and German law a novelty examination must be undertaken before grant can commence, but the search can be delayed for a stipulated period. The early disclosure dates are significant in that the full text of a specification may be available as a Convention application in French or German, for instance, while the text of the original application is still undisclosed in the country in which it was filed, *eg* the United Kingdom or the United States. By scanning the patent specifications published in the early publication countries, it is possible for an individual organisation to gain early warning of its competitors' pending British and American applications. Belgian applications are laid open to public inspection at the Belgian patent office three months after the application is received, but printed specifications are not available until two years after the original date of filing. Published abstracts of these applications are, however, given in the Derwent publication *Belgian patents report* three weeks after filing, and these summaries can be extremely valuable alertings of a competitor's British and American applications. Dutch applications are laid open to public inspection eighteen months after filing, and the published abstracts of these documents given in Derwent's and other publications such as *Chemical abstracts* are again valuable alerting services.

FRANCE

The French patent system can be traced back to the early sixteenth century, although formal patent laws were not established until 1791. A new French patent law has been in force since January 1 1969. As under the former system, French patents are granted for a period of twenty years from the date of application. The subject of the application must have an industrial character, must be novel and must not obviously derive from the prior art. Patents for medicaments are obtainable under the new law. The subject of a French application must not have been disclosed to the public elsewhere in the world by means of a written or verbal description or by public use. The 1969 law retains the system of compulsory licences; these may be applied for if patents are not worked within three years of the grant or after any three consecutive years of non-working. No opposition procedures exist under French law, although when the specifications are laid open to public inspection, eighteen months after the application date, interested third parties may draw attention to anticipatory documents. The former résumé which appeared at the end of the specification is now replaced by claims. The German form of claim is recommended: claim 1 should consist of a preamble summarising the state of the art and should then present a clause specifying novelty and contrasting with the preamble. Applications for patents are examined for novelty but this examination may be deferred for up to two years (unless the subject of the application relates to a medicament) if the applicant so wishes. During this two-year period an application for a patent can be converted into an application for a ' certificate of utility '. These grants which relate to less important inventions are not examined for novelty and remain in force for only six years. Provision exists under the new law for 'certificates of addition'. These can be attached to either a patent or a certificate of utility and have the same expiry date as the parent grant.

BELGIUM

Belgian patent specifications are issued with only formal examination and are granted within a month of filing. Under Belgian patent law, to be novel the invention must not have been described in any published document and must not have been commercially worked by a third party in Belgium. There is no provision under Belgian patent law for compulsory licences. Three months after the grant a typewritten copy of the specification may be consulted in the Belgian Patent Office, but printed specifications are not available until two years after the filing date. There is no formal opposition procedure under Belgian patent law and Belgian patents are granted for twenty years from the date of filing.

NETHERLANDS

The principle of deferred examination, proposed in the draft European convention, was introduced into Dutch patent procedure in January 1964. The applications are laid open for public inspection eighteen months after the filing date. A rigorous novelty search is made by an examiner only at the request of the applicant, or by the intervention of a third party and on payment of a fee. Under Dutch patent law novelty can be claimed only when there has not existed sufficient public knowledge or description to enable a person skilled in the art to manufacture the subject of an application. Applications for patents can only be maintained by the payment of an annual renewal fee. The granting procedure can be commenced by the Dutch Patent Office only after a report has been made by an examiner. If no search has been undertaken within seven years of the publication of the specification, the patent application will lapse. After the application is accepted by the Dutch Patent Office a printed specification is made available for public inspection. An application may be opposed within four months of the publication of the printed specification. Compulsory licences can be requested if the invention has not been worked within three years of the grant and such working would be in the public interest. A Dutch patent will remain in force no longer than twenty years from the date of filing the application. Patents of addition may be obtained in the Netherlands.

FEDERAL GERMAN REPUBLIC

All German patent specifications must be subjected to a rigorous search for novelty before grant. Under German patent law, the invention must not have been described in a document published anywhere within the last hundred years and it must not have been worked commercially in Germany. As each German patent specification must contain a detailed survey of the prior art, these documents are particularly valuable sources of technical information. Compulsory licences may be requested any time after publication of the patent specification if working would be in the public interest.

Deferred examination procedure was introduced under the German patent law of 1968. Under the new Act applications are laid open to public inspection and published as 'offenlegungs-schriften' within eighteen months of the initial date of filing. After an application has been laid open to the public any individual may draw the attention of the German Patent Office to information relevant to the subject of the application and this will be added to the public file. The applicant or a third party may request a novelty search anytime within a period of seven years after the application has been made publicly available. Annual renewal fees are payable to maintain an unexamined application and if no examination is requested within the seven year term the application will lapse. The examination procedure concludes with the acceptance or rejection of the application for publication. After the examination, published applications are referred to as 'auslegeschriften', when the patents are sealed the specifications are again published as 'patentschriften'. Opposition proceedings can be undertaken within three months of the publication of the patentschriften. The former restrictions on the patenting of chemical compounds, foodstuffs and pharmaceutical compounds have been removed under the 1968 Act.

A German patent is granted for eighteen years from the date of application. German Patent Office law also covers the granting of 'gebrauchsmusters', or 'petty patents', which relate to minor inventions which show improvements over known inventions or which lack the degree of novelty needed to sustain a full patent application. No examination is given to 'gebrauchsmusters' but the grant, which is for three years only, will be annulled if prior publication can be proved.

ITALY

Italian patents are granted for a term of fifteen years from the date of application. Details of the invention must not have been published anywhere and it must not have been publicly worked anywhere. Applications are not examined for novelty and there is no provision for formal opposition under Italian patent law. Provisions for compulsory licences were introduced under a Presidential decree in 1968. Italy is one of the few countries which will not grant patents for manufacturing pharmaceuticals.

JAPAN

Japanese patents are granted for a period of fifteen years from the publication of the patent specification, which is examined for novelty. Under Japanese patent law requirements for novelty are that the subject of the application must not have been described in a printed publication, either in Japan or abroad, or publicly known or worked in Japan. Compulsory licences may be requested if the invention is not worked after any period of three consecutive years. Under the revised law of 1971 all applications must be laid open to public inspection eighteen months after the application. Deferred examination procedure similar to that of the Netherlands requires a novelty search to be made within seven years of the application.

USSR

Three types of invention or discovery are recognised:

1 Discoveries, which include only the discoveries of previously unknown natural laws or phenomena; consequently, few grants are made within this category.

2 Rationalisation proposals, which cover ideas which are not inventive but which relate to improvements in existing industrial techniques and methods of production.

3 Inventions, which must show novelty and utility. Either a patent or an author's certificate may be granted for an invention.

Although patents which grant exclusive rights to the inventor are theoretically available in the USSR, the patent system and its inherent connotations of personal property are contrary to the

27

principles of Soviet socialism and only about one per cent of inventions are patented by Soviet citizens. The vast majority of inventions are covered by an author's certificate in which the State reserves the right to use the invention, but grants the inventor financial compensation if it is used. Author's certificates and patents are both subjected to examination for novelty and utility, but no provision is made for an opposition period. Any non-Russian person or organisation can apply for an author's certificate and if this is granted, adequate financial compensation will be paid if the invention is utilised by the State. The USSR was the last of the countries listed above to ratify the *International convention,* the date of its ratification being 1 July 1965.

EUROPEAN PATENT

The present variations in the patent laws of European countries make it a difficult undertaking to obtain protection for the same invention in different countries. To overcome these hazards there have been recent proposals for a ' European ' patent which would extend, initially at least, to the common market countries. The proposed patent laws do not seek to supersede the national patent systems immediately, but aim to coexist with them for an interim period of ten years.

The idea of a European or world patent is not new. In 1909 a German, Raymond du Bois, put forward the idea of a worldwide patent. During the first world war the allies examined the feasibility of a multinational patent, and during the second world war the possibility of a European patent was discussed in Germany. The foundations for a European patent have been laid by the Council of Europe, under whose auspices two conventions have been ratified. In 1953 twelve nations signed the *European convention relating to formalities required for patent application.* This covered procedure—the layout of the patent application form, the number of copies of the specification to be tendered, etc. A further convention dealing with patent classification was signed by the same countries the following year. There is provision for international convention countries to ratify both of these conventions, even if they are not members of the Council of Europe.

The council has drafted a convention on the harmonisation of patent law which aims to standardise patent law in

28

individual countries. Divergent practice causes difficulties when an inventor seeks patent protection in several countries—at present, typewritten disclosure of an invention in the United States presents no bar to patenting in the United Kingdom because the typewritten document is presumed not available in that country, but the same disclosure would disqualify the application in Western Germany where publication in any form in any country is a bar to a grant of a patent. If, as has been suggested, the application for a European patent could also be converted into applications for patents in contracting states, the prior harmonisation of existing national patent laws would be essential.

The creation of a European Patent Office to issue European patents was proposed by the Haertel committee of the European Economic Community in a draft convention which was published in 1962. This action has been deemed necessary because few of the existing national patent systems are now capable of coping efficiently with the increasing volume of patent applications. The European patent law would be based on the principles of deferred examination and intervention procedure, as typified in the new Dutch law, and would ensure early disclosure, thus facilitating the rapid dissemination of information.

After the initial draft was issued in 1962, the problems of implementing the convention were considered but little agreement was reached and work was suspended for three years. The idea of the European patent was revitalised in November 1968 when the French government proposed that work should be resumed, and suggested that the new patent system should aim at a wider territorial setting than that of the six. A body of specialists representing nineteen countries, again under the direction of Dr Haertel, has since produced a revised and extended draft which should be published before the end of 1971. The development of the European patent will be coordinated with the Patent Cooperation Treaty (PCT).

PATENT COOPERATION TREATY
A momentous event in international patent activity has been the adoption of the PCT. The idea for the treaty was first mooted at the 1966 Geneva meeting of the executive committee of the International Union for the Protection of Industrial Property, when a

2

resolution was adopted that the Director of the International Bureau of Intellectual Property (BIRPI) should undertake a study which would lead to a reduction of the duplication of effort involved in obtaining a patent for the same invention in a number of different countries. A final draft of a treaty was produced in 1969 and this was unanimously adopted in Washington in June 1970 by twenty nations. It is now awaiting ratification by the individual countries.

Under the PCT an applicant will be able to obtain an international search to test the novelty of his invention and then nominate a selected list of countries in which he would wish to hold patents. BIRPI will be responsible for the distribution of the patent application to countries named by the inventor. The contracting states will form the International Patent Cooperation Union, which will operate through an international bureau. Existing national patent offices will cooperate with the bureau and will receive international patent applications from their nationals. After a formal examination a copy of the application will be sent to the international bureau and simultaneously to an international searching agency. It is envisaged that there should be five such agencies—the International Patent Institute at the Hague and probably the national patent offices of West Germany, Japan, the USSR and the United States. Within three months of his application a search report consisting of a list of citations will be sent to the applicant, who can then request a preliminary examination as to whether his invention is not obvious and industrially applicable. The examination report will finally be sent to each of the countries nominated by the applicant, and grant in each country will be a decision for the individual national patent office. The period of grace allowed to the inventor between the time of filing in his own country and in other countries will be increased from twelve to twenty months. This extended period should assist applicants to assess on a sounder basis the potential of their inventions before deciding to incur the additional expenses involved in patenting in other countries.

There has recently been progress towards a Nordic patent. In 1967 Denmark, Finland, Norway and Sweden adopted new patent laws which are substantially the same in each country, and work is progressing to the stage where it will be possible to obtain one patent covering all four countries.

The International Patent Institute (IIB) (Nieuwe Parklaan 97, the Hague, Netherlands) is a non-profitmaking institute founded in 1947 on behalf of the governments of Belgium, France, Luxembourg and the Netherlands. Since its foundation, other governments, including that of the United Kingdom, have subscribed to the IIB. The principal work of the IIB is the examination of patent applications on behalf of nationals of its member countries. Novelty searches, which are undertaken on payment of a fee, are made of the specifications of Belgium, France, Germany (FDR), Luxembourg, Netherlands, Switzerland, the United Kingdom, and the USA, in addition to extensive collections of monograph, journal and abstracts literature. The search report gives bibliographical details of the relevant documents identified by the search followed by a summary of their contents. Attention is drawn to any significant technical features which a reference might have in common with the invention examined.

Chapter three: Patents as a source of technical information

*T*he front page of a British patent specification is headed by the specification number. Immediately following this are the filing details. From specification 1,200,001 onwards these details have been given in the standard order advocated by the Committee for International Co-operation in Information Retrieval among Examining Patent Offices (ICIREPAT). Each item of information is followed by an indicating digit, thus the application number is tagged (21), the filing date (22) and the date of filing the complete specification (23). This notation has been recommended to all national patent offices. Its general adoption will bring great benefits to those who work in the field of patent literature. Items of information for indexing and filing purposes could be extracted from patent specifications by persons who had no knowledge of the languages of the specifications or of the patent procedures of the countries involved. The above items are followed by Convention application (31), filing date (32), convention country (33), date of publication of the complete specification (45), international classification number (51), British Patent Office classification marks (index at acceptance) (52), title of the specification (54), name of the applicant (71) and name of the inventor (72). The official titles of British patent specifications are often too broad to be informative and consequently they are of little use in documentation.

The text of the specification can be divided into three parts:

1 *The introductory matter,* commencing with a formal petition from the applicant(s) requesting that a patent be granted to him. This is usually followed by a statement of the field of technology to which the invention relates, a survey of the prior art with examples of previous attempts to solve the problems which the inventor has overcome, and a statement of the objects of the invention.

2 *The body of the specification* commences with the ' statement of invention ' or ' consistory clause ' which describes the essential features of the invention, very often in the language of the main claim; this is followed by a comprehensive description of the invention and the drawings. The statement of invention often starts with the phrase ' according to this invention . . .'.

3 *The claims* which complete the specification are essentially legal in phraseology and define the monopolies claimed by the inventor. British specifications will usually list the main claim first, all following claims being qualifications of the previous one. In the ancillary claims the broad statements made in the first claim are refined. Each claim is autonomous so that if claim one is shown to be invalid because the patentee has been too ambitious in his aspirations it can be replaced by the more specific succeeding claims into which further limitations are written.

The patent specifications of other countries normally follow the layout of the British specification, although in US specifications the main claim is often not the first one listed. US claims are usually more numerous and narrower in scope than British claims.

The text of each US specification is preceded by an ' abstract of the disclosure ' provided by the applicant. Aslib have published a useful *Guide to foreign-language printed patents and applications* which gives examples of the headings of actual specifications from twenty one countries accompanied by English language translations.[1] The publication attempts to be a ' guide for those employed in industry and elsewhere to the extracts from the often perplexing mass of figures and dates on printed patent applications and specifications of those which are of importance in establishing the status of the application or patent, *ie* such details as priority date, filing date, publishing date, etc '.

The patent offices of the United States and Germany print at the end of each specification they publish a list of other related specifications which were consulted by the examiner during his search. Details of patents specifications consulted by the British examiners during a patent novelty search can be obtained from the Patent Office on filing a standard form with a five pence fee.

33

The potentialities of patents as a source of information are little appreciated; indeed, patents are perhaps the most undervalued of all sources of information. The files at the US Patent Office library are probably the largest single body of technical information anywhere in the world—it has been estimated that they contain one million pages of text on chemistry alone.[2] Newman maintains that these files, in conjunction with the US Patent Office classification scheme, can do for much of the electrical and mechanical fields what *Chemical abstracts* does for the chemical fields. He insists that scientists and engineers can no longer afford to ignore this vast well of information.[3]

' PATENTESE '

One reason for the neglect of patent specifications as a source of information is the forbidding style in which many are written. The reader often wonders whether the information is being concealed rather than revealed. Although Judge Learned Hand was speaking specifically about American practice when he referred to ' a barrage of claims ' which are substantially duplicates, and criticised it as ' that surfeit of verbiage which has long been the curse of patent practice, and has done much to discredit it ',[4] most patent systems suffer from ' patentese '. Patents that are easier to read and understand are needed. It has been suggested that ' some attorneys deliberately practise obscurantism in the belief that they can robe an ordinary offspring of mechanical skill in the clothes of invention ',[4] such deception, which is extremely rare, will usually be detected by the examiner or by an opposing attorney if a law suit ensues. US patent 2,938,636 is a rare example of a patent which has been drafted in a functional, jargon-free style, and which is guided by headings which enable the reader to extract information quickly. The claims, which are paragraphed for easy reference, are immediately preceded by a conclusion which summarises the objects of the invention. The survey of the information needs of physicists and chemists undertaken on behalf of the Advisory Council on Scientific Policy[5] suggested that a reason why patents were not more widely used as sources of information was that the ' scientific layman finds them difficult to use '.

The information contained in patent specifications can be used in several ways:

1 If an invention which a firm wishes to use has been patented, the interested firm can approach the patentee for a licence to use the invention. If the validity of the patent is in question, and the interested party can uncover a prior publication which completely anticipates the invention, he can challenge the patent in a court of law. If his case against the patent is successful it will be revoked and he can use the invention without payment of royalties.

2 The claims of a patent specification are very often narrower than the information disclosed in the body of the patent. Any information not covered by claims is freely available for public use.

3 Ideas disclosed in patent specifications relating to one field can stimulate developments in a completely different field. An engineer working on cigarette making machines was studying the problems of designing a mechanism which would automatically splice an exhausted reel of paper to a full reel as soon as it had run out without stopping the machine; no similar work had been done on cigarette making machines but the printing industry had faced an analogous problem. A survey of patent specifications on printing machines gave him the idea for a solution to the problem.[1]

4 A patent is very often allowed to expire or lapse without exploitation. The invention may be patented before a suitable material is available for its development, or the industry may still lack the technical links for its economic use. It may be that the invention cannot find a promoter with vision. Dr Robert Geddes' rocket patent, USP 1,980,266, was rejected by the US government and some years later the Germans utilised the basis of Geddes' patent to produce the V2. Fermi's basic patent on nuclear transmutation, USP 2,205,634, was offered to the same government for a nominal sum but rejected; it later cost the US government some $500,000 in royalties to utilise the patent.[6]

5 Taken collectively, patent specifications are a comprehensive history of the technical development of a nation. The US sub-class files of specifications are ' the most comprehensive textbooks. They show historical development, the reasons for failure, the ways to overcome it, the new trends in development, they

show actually millions of technical solutions, most of them free to public use, some not yet . . . the development engineer should have this textbook always at his fingertips, ready for consultation . . . freeing his mind for the new problems, avoiding the recreation of old inventions . . . making him ready for new approaches by showing the multitude of old efforts '.[7]

6 A topical application of patent specifications is their role in technical forecasting. They can be of relevance to the planning expert who needs to predict and anticipate developments in technology. Descriptions of inventions are often published which may have long-term implications although their immediate application may not be possible because of the gaps in the technology. R P Vcerasnij has observed ' since patents record each step in man's growing understanding of the world around him they can be most successfully used not only to plot the trends of development in the past and present, but also to anticipate the future '.[8]

In Great Britain and the United States, class arrangement of patent specifications is available only in the Patent Office libraries. In Germany many large city libraries make available sub-class arrangements of German specifications which group all patent specifications on related subjects together, thus making prior art surveys a simple matter. In too many organisations the patent is considered merely as a defensive document and patent literature is all too seldom used as it was intended—to liberate ideas for the public good. The engineer who knows his prior art only through textbooks and journals very often misses solutions to technical problems which may be freely available for public use.

THE PATENTS OFFICER

Only the largest firms will employ ' tame agents ', but many medium sized organisations appoint a patents officer to look after their patent activities. The officer does not draft the specifications, as this is the agent's function, but he will be responsible for all liaison between the company and the patent agent it contracts to prosecute its patent applications. The patents officer's tasks include: literature searches to establish the novelty of an invention and to clear the way for patenting; the payment of renewal fees on the patents the company holds (the patent agent will send out renewal notices of impending payment of fees for the patents he has prosecuted for his clients, but the final responsibility for payment is that of the company); watching the patent literature for specifications, both British and foreign, of relevance to the activities of the organisation; translation of foreign patent specifications—as legal phraseology is used in foreign patent specifications it is essential that this work should be undertaken by a person with a sound knowledge of patent terminology in addition to a knowledge of foreign languages; the maintenance of indexes to the patent specifications held by the organisation; information retrieval from patent specifications; and the preparation of a patent abstracting bulletin. The patents officer must have a broad knowledge of the technical background to his organisation's activities, a good knowledge of patent procedure and as wide a knowledge of foreign languages as possible. In many smaller firms the functions of the patents officer may be carried out by the information officer or the technical librarian.

PATENT ABSTRACTING

Most large industrial organisations issue a patents abstracting bulletin to keep their scientists and technologists informed of patent activity within their fields of interest. The officer undertaking the abstracting for this publication will concentrate on the data given in the broadest claim of the specification; it is this claim which will form the basis of the abstract, although information on the uses, applications, or alternative methods of carrying out the invention may be extracted from the body of the specification to supplement the information given in the claim. The abstractor will also usually substitute a suitably descriptive title to replace the inadequate official title of the specification and to introduce his abstract. Filing details of the specification must also be included to cater for company officers who are concerned with the legal implications of patents.

FILING OF PATENTS

Specifications are invariably filed first of all by country of origin, then by specification number, for it is by this number that they will be requested and cited. Libraries which do not acquire complete sets of the specifications published by any one country will file their specifications loose in box files or transfer cases, each box being labelled with the first and last numbers it contains. An added aid to the searcher is to attach a typed list of the specifica-

2*

tion numbers included in each box on its inside cover to indicate immediately whether a particular specification is held. Public libraries maintaining complete files of British patent specifications will bind these documents into volumes containing one hundred specifications. The British Patent Office issues title pages for every series of one hundred specifications it publishes.

INFORMATION RETRIEVAL

The highly complex nature of the subject content of patent specifications will render the subject organisation of this material by conventional systems impracticable. Post co-ordinate or uniterm indexing is now widely adopted to retrieve information from patent specifications. This method of indexing releases the indexer from the task of expressing the numerous relationships which can exist between the component parts of a complex subject. His function is reduced to isolating the component parts of the subject and to posting the terms which express them onto aspect or uniterm cards. The co-ordination of the component parts of a complex subject on which a search is required is carried out at the moment of searching by comparing the relevant aspect cards either manually or mechanically. This method of organising information lends itself admirably to computerisation and in recent years several firms have developed sophisticated computer-based systems which utilise concept coordination to retrieve information from the patent literature. Post co-ordinate indexing is now also widely used for retrieving information from technical report literature,[9, 10] the subject nature of which is usually even more complex than that of patent specifications.

Some libraries maintain an index of patentees in addition to their subject indexes. This index will include name of patentee(s), title of the specification, specification number and filing details. If the cards also include abstracts of specifications, an efficient method of producing a bulletin is readily available. The abstract cards can be arranged to form a ' mock-up ' page by securing them on a white Bristol board and then producing an offset litho master by means of xerography. If the abstract cards in use are too large to produce a page of convenient size it is possible to reduce the size of the xerographic master by photographic means. After the processing has been completed, the cards are dismounted and returned to the index, the required number of

38

pages are run off on the offset machine from each master, collated and stapled together to form the bulletin.

REFERENCES
1 Finlay, A F: *Guide to foreign-language printed patents and applications.* Aslib, 1969, iv.
2 Willson, R C: *Chemical engineering* 71 February 3 1964 79-84.
3 Newman, S M: *Journal of the Patent Office Society* 43 (6) June 1961 424.
4 Lutz, K B: *Journal of the Patent Office Society* 42 (8) August 1960 568.
5 *Journal of documentation* 21 (2) June 1965 83-90.
6 Puishes, A: *Mechanical engineering* 83 (8) August 1961 47-9.
7 Wimniesperg, H von: *Journal of the Patent Office Society* 42 (3) March 1960 172-3.
8 Vcerasnij, R P: *Unesco bulletin for libraries* 23 (5) September-October 1969 236.
9 Grosch, A N: *Special libraries* 56 (5) May-June 1965 303-11.
10 Johnson, A: *Journal of documentation* 15 (3) September 1959 146-55.

Chapter four: Bibliographical control of patent literature—official sources

*B*etween 320,000 and 340,000 patent specifications are published throughout the world each year. The United States is the largest publisher of specifications, in the fiscal year 1969/70 78,000 documents were issued and in the fiscal years 1971 and 1972 the projected figures are 83,000 and 80,000 respectively. France and Great Britain, each with around 40,000, are the next largest publishers—these countries have doubled their numbers of specifications published per year since 1939. Italy issues about 35,000 specifications a year, Japan 28,000, Canada 25,000, Germany (FDR) 21,000, Belgium 17,000 while Switzerland, Spain, Australia, Brazil, Austria and India all usually publish more than 5,000 specifications a year.

Although they do not publish as many specifications as the leading patent publishing countries, Switzerland, Belgium and Czechoslovakia publish more patents per head of their populations than any other countries. Each December issue of the journal *Industrial property* (an English translation of *Le propriété industrielle*) gives details of the number of patents issued during the previous year by a number of countries who are parties to the international convention.

EQUIVALENT PATENTS

The information worker must realise that not all the patent specifications published throughout the world each week on his subject will yield new information. Not every one of the thousands of specifications published each week describes a unique invention, for an invention may be patented in a number of countries; a British patentee may seek protection in the United States, France and Japan, thus owning four patents for the same invention. The 340,000 plus specifications mentioned above relate to 120,000 basic patents: each patent is taken out in three countries on the average. As filing abroad can be extremely costly—fees for a single country can amount to £120, not including the patent agent's fee—companies are usually selective in filing applications only in those countries which have the technical facilities to produce the invention and the market to absorb it. Two thirds of all British applications are also filed abroad. In Great Britain an inventor cannot file a foreign application before a British application without obtaining the permission of the Comptroller of Patents. With less significant inventions some companies merely 'disclose' them in foreign countries (*eg* by publishing a description in a technical journal). This does not give them exclusive protection, but prevents any competitor from being granted a patent for the same invention.

Monopoly rights exist only in the country in which the patent is granted; therefore the disclosure in a foreign patent specification may be utilised in the British Isles while the invention is not covered by a British patent. To ascertain whether a British equivalent of a foreign patent specification exists it is necessary first to check the name indexes published by the Patent Office, and available in the Patent Office Library, under the name of the foreign patentee, then to check the filing details of the foreign specification against any possible British equivalents traced through these name indexes. The identification of equivalent patents is often rendered difficult as the name of the applicant may be different due to assignment; in many cases the only method of identifying equivalents is by undertaking a subject search.

The technical information division of the Esso Research and Engineering Co, Linden NJ have approached the problem of equivalent patents by compiling a 'convention-date file', a set of patent abstract cards filed by convention country and sub-filed by convention date. Thus a British patent bearing a US convention date would be filed under United States, then by its application date. Anyone checking a patent specification for corresponding cases simply has to consult the file under the convention date of the specification in question; when an abstract has been indentified as equivalent or corresponding, the country and specification number of the equivalent patent is posted onto the abstract card in the 'convention-date file'.[1]

PATENT JOURNALS
The various patent offices of the world publish journals or gazettes to keep industry informed of patent activity within the

country. These tools, together with unofficial sources such as patent information services and patent abstracting journals, enable patent agents, librarians and information officers to select and obtain those patent specifications which are relevant to the work of the organisations which they serve. Abstracts or notices of new patent specifications also appear in many of the scientific and technical journals covering particular industries or disciplines.

I OFFICIAL SOURCES : GREAT BRITAIN

Notification of forthcoming British specifications is given in the *Official journal (patents)* (OJ) six weeks before their publication. Each week this journal gives a list of complete specifications accepted, in specification number order. For each specification the following information is given: seven figure serial number (specification number); application number; name of patentee(s); brief title; date of filing the complete specification; date of filing the provisional specification; or the convention country and filing date of the basic patent in cases of convention application; main British and the international classification marks. The journal gives a weekly list of applicants for patents, each entry including the name of the applicant; a brief title of the specification (which never contains any vital information thus preventing premature disclosure of the specific nature of the invention); whether the application was accompanied by a complete or provisional specification; convention country and filing date (if appropriate). A list of application numbers for which patents have been granted is included enabling interested parties to follow the progress of applications which have been previously noted. Other information given in the OJ includes a weekly name index of patentees, a subject index of complete specifications accepted based on the *Classification key* and a list of specification numbers for which letters patent have been granted.

The *Abridgments of specifications* are the main tools for searching for British patent specifications on a particular subject; these abridgments are based on the *Classification key*. Each abridgment gives the specification number; title of the specification; name of the patentee; date of application; the Patent Office classification headings under which the specification has been indexed and a detailed illustrated abstract of the specification which is written by the examiner at the time he undertakes the novelty search. The abridgments are now issued within one week of the publication of the corresponding specifications. It should be noted that several abridgments for the same specification will appear in various divisions of the abridgments if the examiner has allocated several press marks (classification headings) to a specification which covers several subjects. In such a case each abridgment will have varying subject matter, each describing only those features of the invention which correspond to the press mark under which the abridgment appears. References are given at the end of each abridgment to enable the searcher to ascertain under which other press marks the specification has been abridged.

The system of classification of British patent specifications was changed from the publication of BP 940,001. The *Classification key* is now divided into eight sections lettered A-H; each section is then divided into two or more divisions, making forty in all; each division is again subdivided into a number of headings, four hundred and nine in all. Section F of the classification covers mechanics, lighting and heating and is divided into four divisions :

F1   Prime movers, pumps

F2   Machine elements

F3   Armaments, projectiles

F4   Lighting, heating, cooling, drying.

Division F4 is subdivided into fifteen headings :

F4A   Steam generation

F4B   Furnaces

F4F   Lighters

F4G   Drying

F4H   Refrigeration

F4J   Chimneys

F4K   Direct contact heating of fluids

F4P   Gas storing, liquefying

F4R   Lamps

F4S   Indirect heating and cooling

F4T   Burners

F4U   Heating and cooling systems

F4V   Ventilation and air conditioning

F4W Stoves and fireplaces

F4X Gas distribution, washing-boilers.

Each of these headings is again very specifically subdivided. The scheme is not constructed according to any theoretical rules, as are bibliographical classifications, but is essentially an arbitrary grouping of topics which has been developed chiefly as an aid to the Patent Office examiners in their novelty searches. Before a specification is accepted it is classified by an examiner under several headings taken from the *Classification key;* after this classification, the examiner will search all prior British patents for the previous fifty years which bear the headings which he has assigned to the patent specification under scrutiny, to determine whether the subject of the application is novel. The sequence of the divisions of the *Classification key* is similar to the arrangement of the *International patent classification,* whose numbers three quarters of all patent publishing countries assign to their specifications. A card index of patents issued after 1 January 1956 in Belgium, France, Great Britain, Luxembourg, Switzerland and the USA, maintained at the Institut National de la Propriété Industrielle in Paris, is arranged by subject according to the IPC.

The abridgments are issued weekly in sheet form inside twenty five groups, each group comprising one or more of the forty divisions of the *Classification key.* When 25,000 specifications have been issued the Patent Office supplies subscribers to the abridgments service with title pages for the various groups and also the relevant sections of the *Classification key* for the groups together with *Subject matter index* which enables a searcher to ascertain which specifications have been allocated to a specific press-mark inside a particular heading. The abridgments issued in sheet form can then be bound up into volumes, each covering a series of 25,000 specifications, for retrospective search.

The reference index to the classification key is in three sections. The first is an alphabetical list of over five thousand catchwords which will indicate to a searcher at which heading the subject he is interested in has been classified, and therefore which group of the abridgments he must consult; the second part outlines the structure of the new *Classification key;* the third part defines which subjects are contained under the various headings and gives scope notes to indicate what is not included in the heading.

44

A searcher for specifications dealing with, for example, electric battery lamps would first consult the alphabetical catchword index to the *Classification key* to determine under which division and heading this subject had been classified. The index gives the following entry:

Lamp(s)
   electric lamps
     kinds
       battery lamps   F4R

therefore by consulting the abridgment volumes covering the divisions F3-F4 and by then examining the *Classification key* for abridgment group F3-F4 the searcher would be able to determine the specific press mark under heading R which subsumes the subject ' electric battery lamps '. Reference to the *Subject matter index,* a component part of each abridgment volume, would then identify any relevant specifications. By using this approach he could identify all specifications published on the subject since 1963 (BP 940,001). To extend the search beyond this date he would need to consult the *Backward concordance* to the *Classification key,* which is included in the abridgment volumes published in 1964 and which translates post-1963 divisions and headings into their pre-1963 equivalents. This will indicate that abridgment group XI and abridgment class 75 are the relevant pre-1963 sections. By scanning these abridgment volumes the searcher can survey all specifications published since 1855. A useful aid to the retrospective search which is available in some of the larger public technical libraries is the *Fifty years subject index, 1861-1910.*

The unillustrated abridgments for specifications issued from 1617 to 1876 were arranged chronologically within 100 subject classes with name and subject indexes appended to each volume. Each volume usually commenced with a useful résumé of the development of the art, but some volumes did not cover every specification issued within the class, and the volumes for some classes terminated at 1867. The *Illustrated abridgments of specifications* published from 1855 to 1908 were arranged into 146 classes with nine volumes to each class, from 1909 to 1930 into 271 classes with four volumes to each class, and from 1931 to 1963 into 46 classes with each class comprising of a set of volumes covering successive ranges of 20,000 consecutively published specifications.

The index to names of applicants in connection with published complete specifications enables any searcher to find out which specifications have been granted to an individual or organisation. Each volume covers a series of 25,000 specifications and each entry gives the names of the applicant, the brief title of the specification and the specification number.

The divisional allotment index to abridgments of specifications, which covers similar series of specifications to the above, indicates in which divisions of the *Classification key* an abridgment of a specification has been published.

No other patent office provides such efficient searching tools for their specifications as the British Patent Office. A pamphlet entitled *Searching British patent literature* which gives detailed guidance on the use of the *Classification key* and the abridgments is available from the office on request. The abstracts contained in the gazettes of most other countries give only brief details of the main claim, thus necessitating reference to the original specification in many cases. Although some of the gazettes are classified, they are inadequate as searching tools because their abstracts are not issued in separately cumulated groups as with the abridgments. Searching foreign patent literature is then a much more time-consuming operation than searching British specifications.

NATIONAL REFERENCE LIBRARY FOR SCIENCE AND INVENTION (NRLSI): SEARCHING FACILITIES

Although complete sets of Patent Office publications are distributed to sixteen provincial public libraries, and it is possible to undertake a novelty search in these libraries, searching is facilitated in the NRLSI by the provision of a series of specialised indexes which are not available elsewhere.

Sheffield City Libraries have recently issued a useful guide entitled *Patent holdings in British public libraries* (third series), 1970, which lists the holdings of nineteen countries. Under the name of each country the holdings of twenty libraries are given with the dates of commencement of their files. This gratis publication is available on application to the libraries of Commerce, science and technology.

The NRLSI indexes are:

1 A system of class boxes enabling a searcher to survey the recent specifications published on a particular topic. As each successive week's specifications are received in the library one set is filed in a sequence of boxes which is arranged in the same classified order as the specifications will be indexed in the abridgment volumes. These specifications are withdrawn from the class boxes after a period of six weeks.

2 Pre-abridgment specifications subject matter index: an automatically produced listing of specifications which have not yet been included in abridgment volumes, which cumulates the subject matter index from the OJ.

3 A card index of the names of applicants for patents granted since the publication of the most recent *Index to names . . .;* as with the abridgments the cards are withdrawn on publication of the appropriate name index.

4 A card index of the names of applicants for patents during the current year and for at least the previous six years.

5 The *Application register* (red series), available from 1914, which relates the application number to the final specification number assigned on acceptance, or indicates if the application was abandoned or declared void.

6 The *Register of stages of progress* (green series), available from 1914, which lists the patents in numerical order, indicating the dates of completing the principal stages in prosecution from application to sealing.

7 The *Register of renewal fees* (maroon series), available from 1914, which lists patents in specification number order and indicates whether the annual renewal fees required to keep the patent in force have been paid.

DEPOSIT ACCOUNTS

An individual who wishes to obtain copies of all British patent specifications on one particular subject can open a deposit account with the Patent Office. The sales branch will supply on a weekly basis all specifications allocated to any heading from the *Classification key*. The minimum amount that the Patent Office will accept to open an account is £4. As the current price of a British specification is 25p, anyone wishing to take advantage of the *Selected specifications service* is advised to open an account with a much larger sum. A similar but more restricted service which facilitates the scanning of the abridgments or the actual specifications is the provision of weekly tabulations of specification numbers relating to a specific topic. The cost of the service is 5p per printed sheet.

FILE LISTS

Subject matter file lists, which list all specification numbers allocated to any heading from the *Classification key* over a period of fifty years, are also available from the Patent Office. The required features must be indicated as combinations of headings from the *Key*. File lists are only produced to fulfil specific orders and are not available for reference in the Patent Office library. To undertake a patent novelty search without the tedious task of searching the abridgment volumes through their own indexes, the searcher can discover under which heading the subject matter of his invention may fall and then obtain file lists from the Patent Office for these headings. The abridgments can then be consulted in association with the file list and the searcher need only refer to the actual specification when an abridgment indicates its relevance.

MECHANISATION OF PATENT SEARCHING

The Patent Office file lists are produced by mechanical card sorting devices. A total of over 6,000,000 individual cards, sorted into 45,000 decks, one for each term in the classification, are held by the Patent Office for searching purposes. The patent offices of several countries are examining the possibilities of the mechanisation of patent searching. The *Committee for International Co-operation in Information Retrieval among Examining Patent Offices* (ICIREPAT) was established in 1961, and now includes in its membership eighteen nations and two treaty organisations. Its main purpose is to foster international co-operative research into information retrieval for patent novelty searches. ICIREPATS's programme for the shared indexing of patent specifications by national patent offices has been usefully summarised by Pfeffer.[2] ICIREPAT is also concerned with the improvement of patent classification schemes and the problems created by equivalent patents.

The United States Patent Office is now developing a computerised system, HAYSTAQ, which is designed for searching chemical patent specifications. If the system is successful it is hoped that HAYSTAQ techniques can be applied to other subject fields.[3]

The office is also making available at economic rates duplicate copies of all US specifications on microfilm aperture cards at a reduction ratio of 22:1. The standard punched cards will incorporate a picture frame mounting for a microfilm which will bear an abstract and up to eight pages of the specification. It is estimated that only one such card will be needed for 90 percent of US specifications. In addition to the advantages in space savings which will accrue to the industrial user of patent specifications the development of the microfilm system will relieve storage problems at the patent office in that it will largely eliminate the need for keeping bulk copy stocks of individual specifications in hand to satisfy requests for copies of a particular patent number. Another recent programme is the commencement in 1970 of work on a tape library of US specifications. It is planned that within three years the file will include all current US specifications and ultimately it will embrace the retrospective collections. The tapes will be made generally available for the machine retrieval of patent data.

NRLSI : HOLBORN DIVISION
The NRLSI Holborn Division is based on the former Patent Office Library which was founded in 1855 ' to stimulate developments in the field of invention by making relevant information freely and readily available to all '. For over one hundred years the Patent Office Library contained the largest public open access collection of scientific and technical material in Great Britain. The NRLSI was officially instituted in April 1966 and it is now administered as a department of the British Museum Library. For the development of the NRLSI the stock of the Patent Office Library was augmented from 1963 when staff were appointed to purchase new material and to select relevant literature from the documents received by the British Museum Library under legal deposit regulations. The NRLSI now comprises two divisions. The Holborn Division contains the literature relevant to patentable matters, ie the major part of the chemical, engineering and technological literature together with some material from the life sciences related to these disciplines. The Bayswater Division houses the literature of the life, earth and astronomical sciences.

49

The needs of the layman and undergraduate are not specifically catered for by the NRLSI, as literature is selected on the basis that it shall be useful to the graduate and other professionally qualified readers. Nevertheless, in craft technology where little academic literature is available, material is acquired which will be suitable for those already skilled in their art. The stock of the Holborn Division includes some non-technical subjects such as patent law, which are pervasive of developments in all technical subject fields. The history of science and technology is considered as outside the scope of the library and with the exceptions of patent specifications and bibliographies, only literature published since 1910 is retained at Holborn.

The Holborn Division's stock of more than 450,000 volumes includes British and foreign scientific and technical journals, transactions of learned societies and textbooks and monographs on science and art. Some 10,000 current periodicals are received, in addition to 8,500 other titles which are no longer taken, usually because they have ceased publication. Current bibliographical and abstracting journals are shelved in a separate classified sequence apart from other periodicals, and a subject index of bibliographies contains over 10,000 entries. Separate sequences of pamphlets, classified by subject, and trade literature, arranged alphabetically by manufacturer's name are provided. Trade and commercial directories are shelved adjacent to the trade literature to enable a reader to identify the manufacturers of a certain product and then to scan the library holdings of their trade catalogues. The library maintains a classified catalogue with author and alphabetical subject indexes. Indexes of periodicals, translations held by the library and dictionaries are also provided.

In addition to the above material, the Holborn Division holds extensive runs of the complete or abridged patent specifications of about one hundred countries. The interests of the users of this literature were carefully protected when the Patent Office Library was absorbed by the NRLSI.

The NRLSI Holborn Division publishes *Periodical publications in the National Reference Library for Science and Invention. Pt 3 List of current titles in the Holborn Division* (fourth edition 1970) and also issues a free *Aids to readers series* as follows: 1 *Holdings of translated journals;* 2 *Outline of the classification;* 3 *Abstracting and bibliographical periodicals;* 4 *Literature on patents and trademarks shelved outside the patents sections of the library; 5 Notes on the filing arrangement of the author and periodicals card catalogues; 6 Translations (a guide to their availability); 7 Special holdings of Russian literature; 8 Photocopy service; 9a Holdings of foreign patents and trade mark literature: Germany; 9b Holdings of foreign patent and trade mark literature: Belgium; 10 Literature in the British patents section; 11 Linguistic help; 15 Notes on the author and name catalogues; 19 Translations series.*

Another useful publication of the NRLSI is *Periodicals news lists*, a monthly giving bibliographical information about journals, *eg* changes in titles, amalgamations, and including lists of new journals acquired by the library.

The holdings of the NRLSI must always be available for consultation and it does not therefore make material available on loan; photocopies, however, can be supplied, subject to the Copyright Act, while the reader waits. The service covers all publications held by the library except HMSO publications and British Patent Office publications in print. Photocopies can also be requested by phone or post and a deposit account can be opened for this purpose, the usual cost being 4p per page. Although the NRLSI Holborn Division does not undertake patent novelty searches on behalf of its users, every assistance is given in the use of searching tools and aids such as the *Classification key*. A bibliographical reference service in answer to personal, postal or telephone enquiries is also offered.

2 OFFICIAL SOURCES: UNITED STATES

The United States Patent Office publishes a weekly *Official gazette* (OG) which contains brief summaries of US specifications as they are published each Tuesday. The summaries are listed under three main headings: general and mechanical; chemical; electrical. Within these sections the summaries are arranged in classified order according to a scheme consisting of 300 classes, which are subdivided into some 60,000 sub-classes. To facilitate searching for individual patent specifications, the three classified sequences are arranged so that the specification summaries or abstracts are also in specification number order. For each specification listed the following information is given: specification number; title; name of inventor and assignee; filing details;

number of claims; class and sub-class number; illustration; brief details of the main claim. A searcher can discover the number of the class and sub-class relevant to his interests by consulting the US Patent Office *Manual of classification*. The searcher for US patent specifications on electric battery lamps, by consulting the index would find the entry:

	Class	Sub-Class
Electric and electricity		
Lamps		
Battery lamps	240	10.6

To discover which specifications had been recently published on this subject he would then consult the weekly issues of the OG under the electricity section. It should be noted that only one summary of each specification is given in the OG. As the OG does not cumulate, retrospective searching for US specifications on a particular subject is a tedious operation. The OG, in addition to the summaries of specifications, contains a name index of patentees, a classification of specifications which shows at a glance how many specifications have been allocated to a specific sub-class each week, a geographical index of the place of residence of inventors, but no details of applications for patents.

The United States Patent Office has published an *Index of patents* annually since 1872; this contains two main sequences: an alphabetical list of patentees and a numerical breakdown of specifications by class and sub-class. The *Classification bulletins* available from the United States Patent Office supplement the *Manual of classification* by giving definitions and descriptions of the kinds of specifications included in each class and sub-class, and by referring to other classes which will contain related specifications.

The public search room of the United States Patent Office is the only place in the United States where a complete file of US patent specifications is arranged in subject order; in this room the sub-class bundles of specifications are available on open access. Searchers who are not able to visit the Patent Office can obtain sub-class lists of specification numbers which are prepared by the USPO in response to individual orders. Files of US specifications arranged by specification number are available in twenty two libraries throughout the USA, while a much larger number of libraries regularly receive the OG. The USPO has also issued microfilm lists of the specification numbers of patents issued in each sub-class, which are available in several US public libraries and which will enable a searcher to undertake a novelty search in any library which holds these lists, a copy of the *Manual of classification* and a file of US specifications or a set of the OG. The search room maintains:

1 An alphabetical card index of the names of patentees since 1931, and

2 A card index of every US specification issued, arranged by specification number and giving details of the class and sub-class number under which the specification is filed. This enables a searcher who commences a search with one significant specification number to be guided to the sub-class bundle where other related specifications can be found.

THE UNITED STATES PATENT OFFICE LIBRARY

The scientific library of the USPO contains more than 7,500,000 foreign patent specifications, in addition to a complete set of US patent literature. The library also holds over 120,000 volumes of scientific and technical books and about 80,000 bound volumes of periodicals and abstract journals. In many cases two sets of the patent specifications of a single country are provided, one arranged in specification number order, the other arranged according to the classification scheme adopted by the patent office of the country of origin.

OFFICIAL PUBLICATIONS OF OTHER COUNTRIES

The official journals published by most patent offices contain abstracts of their countries' patent specifications.

The following list shows the journals of the most important patent publishing countries giving 1 title; 2 date of commencement of publication (where ascertainable); 3 frequency; 4 whether classified (C) or arranged by specification number (N); 5 whether the journal includes abstracts of patent specifications or title listing.

AUSTRALIA: *Australian official journal of patents, trade marks and designs;* 1904-; weekly; (N); abstracts.

AUSTRIA: *Oesterreichischer patentblatt;* 1899-; monthly; (C); titles.

BELGIUM: *Recueil des brevets d'invention;* 1854-; monthly; (C); abstracts.

BRAZIL: *Diario oficial;* 1922-; daily; (C); abstracts.

CANADA: *Canadian patent office record;* 1873-; weekly (N); abstracts.

CZECHOSLOVAKIA: *Vestnik uradu pro patenty a vynalezy;* monthly; (C); abstracts.

DENMARK: *Dansk patenttidende;* 1894-; weekly; (N); titles.

FINLAND: *Patenttilehti patenttidning;* monthly; (N); abstracts.

FRANCE: *Bulletin officiel de la propriété industrielle: abrégés descriptifs des brevets d'invention;* 1828-; weekly; (C); abstracts.

GERMANY (WESTERN): *Patentblatt;* 1880; weekly (C); titles *Auszuege aus den patentanmeldungen;* weekly; (C); abstracts. *Auszuege aus den gebrauchsmustern;* weekly; (C); abstracts. (The last two titles are semi-official abstract journals published by Wila Verlag für Wirtschaftswerlung in co-operation with the German Patent Office.)

GERMANY (EASTERN): *Bekanntmachungen des amtes für erfindungs-und patentwesen der Deutschen Demokratischen Republik;* semi-monthly; (C); titles.

HUNGARY: *Szabadalmi kozlony es kozponti vedjegy ertesito;* monthly; (C); abstracts.

INDIA: *Gazette of India: notifications . . . relating to patents and designs;* 1889-; weekly; (N); titles. A separate series of abridgments is also issued.

IRELAND: *Official journal of industrial and commercial property;* 1928-; semi-monthly; (N); abstracts.

ISRAEL: *Patents, designs and trade marks journal;* 1968-; monthly.

ITALY: *Bolletino dei brevetti;* 1902-; semi-monthly; (C); titles.

JAPAN: *Tokkyo koho;* 1886-; irregular; abstracts. *Jitsuyo shinan koho* (Official utility model patent gazette); 1968-; irregular; abstracts.

NETHERLANDS: *Industriele eigendom;* semi-monthly; (C); long titles.

NEW ZEALAND: *New Zealand patent office journal;* 1912-; monthly; (N); abstracts.

NORWAY: *Nordisk tidende for det industrielle rettsvern;* 1886-; weekly; (N); long titles.

PAKISTAN: *Gazette of Pakistan: notifications and notices issued by the patent office;* weekly; (N); abstracts.

POLAND: *Wiadomosci urzedu patentowego;* 1924-; bi-monthly; (C); titles.

54

PORTUGAL: *Boletim da propiedade industrial;* 1896-; monthly; (N); abstracts.

ROMANIA: *Inventii si inovatii;* 1966-; monthly.

SOUTH AFRICA: *Patent journal;* 1948-; weekly; abstracts.

SPAIN: *Boletin oficial de la propiedad industrial;* 1886-; semi-monthly; (N); abstracts.

SWEDEN: *Svensk tidskrift for industriellt rattsskydd;* 1892-; semi-monthly; (N); abstracts.

SWITZERLAND: *Schweizerisches patent und muster und modell-blatt;* 1962-; semi-monthly; (C); titles.

USSR: *Otkrytiya, izobreteniya, promyshlennye obraztsy, tovarnye znaki;* 1924-; semi-monthly; abstracts.

YUGOSLAVIA: *Patenti glasnik;* quarterly; (C); abstracts.

REFERENCES
[1] Kohnke, E L and Lewenz, G F: *Journal of chemical documentation* I (I) 41-3.
[2] Pfeffer, H: IR among examining patent offices. *Special libraries* 59 (5) May/June 1968 330-6.
[3] Marden, E C and Koller, H R: ' Present status of project HAYSTAQ ' in ICIREPAT *Third annual meeting 1963* (Spartan Books, 1964) 163-77.

Chapter five: Bibliographical control of patent literature—non official sources

Several commercial organisations are active in the field of patents documentation. These firms publish patent abstracting journals or undertake watching and reporting on patent literature in specific fields: DERWENT PUBLICATIONS LTD (Rochdale House, 128 Theobalds Road, London WCI) publish the most comprehensive range of journals. Their publications can be divided into three categories: 1 journals covering every patent specification issued within a country; 2 those covering chemical specifications only; 3 the *Central patents index*.

1 Every patent specification within the country: *British patents abstracts* (1951-) is published each Tuesday, the abstracts which appear three weeks after the specifications are laid open to public inspection give full details of the main claim of each specification. Each issue is arranged according to the Derwent patent classification which is also used to classify other Derwent publications. Patents are assigned to one class only and no cross references are given. Each issue contains an alphabetical patentee index giving the specification number, the Derwent classification number and a brief title. A group allotment index in specification number order indicates to which Derwent class the specifications have been assigned.

German patents abstracts (1951-) is published each Thursday and contains abstracts of the *auslegeschriften* four weeks after they are laid open to inspection. The abstracts are in English and the coverage and arrangement is the same as for *British patents abstracts*. Both the above publications are available as chemical editions only as *British patents report* and *German patents report*. The *German patents gazette*, which has been available since October 1968, gives English language abstracts of the one thousand *offenlegunsschriften* published each week prior to examination.

Soviet inventions illustrated (1961-) is published monthly, six weeks after the specifications are first received in Britain. The abstracts are classified as above but the journal is split into three sections, chemical, general and electrical, which may be subscribed to separately.

2 Chemical specifications only: *Belgian patents report* (1955-); *French patents abstracts* (1961-); *Japanese patents report* (1962-); *Netherlands patents report* (1964-); *United States patents report* (1971-). These publications are issued weekly and give English language abstracts arranged by the Derwent classification three to four weeks after the specifications are laid open to inspection.

3 *Central patents index (CPI)*: *CPI* was introduced in January 1970 and now covers chemical and related specifications from the thirteen most significant patent publishing countries. An average of 2,000 basic specifications from Australia, Belgium, Canada, France, West Germany, East Germany, Japan, the Netherlands, South Africa, Switzerland, the United Kingdom, the United States and the USSR are abstracted each week. It is estimated that the remaining patent publishing countries of the world account for only 5 percent of the basic specifications. A series of alerting abstract bulletins including a summary of about one hundred and twenty words is published covering all basic specifications and most equivalents three to four weeks after the publication of the original. The bulletins are available in twelve sections corresponding to the sections of the Derwent classification —Plasdoc (polymers, plastics); Farmdoc (pharmaceuticals); Agdoc (agricultural chemicals); Food, disinfectants, detergents; Chemdoc (general chemicals); Textile, paper; Printing, coating, photographic; Petroleum; Chemical engineering; Nucleonics, explosives, protection; Refractories, glass, ceramics; Metallurgy. Basic abstract journals containing more detailed ' documentation abstracts ' of about four hundred words are also available for the same twelve subject classes. The *CPI* service also offers abstracts on standard IBM cards for manual or machine retrieval, files of the basic specifications of certain countries of 16mm microfilm, and magnetic tape records of patent data for in-house searches using Derwent programs. Hard copy reproductions are provided of every basic patent other than Japanese and Russian for the Farmdoc and Agdoc sections only. Subscribers to *CPI* pay a

'participation fee' dependent on the number of sections and services in which they are interested.

TRANSLATION AND TECHNICAL INFORMATION SERVICES (6 Church End, Banfield, Braintree, Essex) publish several patents abstracting journals. *East European science abstracts* (1965-) is a monthly publication giving complete coverage of all chemical patents issued in Poland, Czechoslovakia and East Germany for native firms. The abstracts are arranged in twenty five groups, *eg* food and drinks; plating; concrete, cement and plaster. For each specification abstracted, the number of words of the original is given and the cost of translation into English. *Paint and resin patents* (1964-) covers the patents of eleven countries (including east European) and is also classified into broad groups. *Organometallic compounds* (1961-) covers the patents of eight countries. *Continental paint and resin news* (1957-) is an abstracting and digest publication covering English language sources in addition to European journals and a selection of patent specifications. TTIS will provide translations into English from most European languages and also undertake searching, watching and abstracting on specific subjects, *eg* all patents from northern Europe on cosmetics.

WATCHINGS

Several services undertake patent watchings, usually notifying their clients by monthly or quarterly reports listing abstracts on the subjects requested. The following firms provide these services:

INVENTIONS INC (Washington 4, DC, USA). Reports on US, British and Canadian patents.

PATENTS RESEARCH AND DOCUMENTATION (London WC1).

POLYSEARCH SERVICE (The Hague, Netherlands). Abstracts on specific subjects from Australian, Canadian, French, German, Mexican, Dutch, South African, US and other countries if requested.

RESEARCH INTELLIGENCE (Philadelphia 2, USA).

ROTHA, FULLFORD, LEOPOLD AND ASSOCIATES (Melbourne, Australia). Japanese patents.

TD NELS (Berlin SW61). Weekly watchings on German patents.

Periodicals which document the patent specifications of a single country or on a specific subject include:

58

Airplane patent digest (Manufacturers Aircraft Association Inc, New York). Weekly.

Austria-patent-reporter/Osterreichischer Patentinhaber und Er-finderverband (Association of Austrian Patentees and Inventors, Freyung 1, Vienna 1010) monthly. Abstracts of Austrian specifications.

Gas Appliance Manufacturers' Association Inc patent digest (1901 North Fort Myer Drive, Arlington, Virginia, USA) monthly. Abstracts of US patent specifications taken from the *Official gazette*.

Graphic arts patent abstracts: US patents related to printing, packaging, paper and photography (Graphic Arts Research Center, Rochester Institute of Technology, New York) monthly.

Inventions for industry, formerly NRDC *Bulletin* (National Research and Development Corporation, 1 Tilney Street, London W1) semi-annually. Abstracts of British patents held by the NRDC for which licences are available to firms, royalty free.

Japan patent news (International Patent Service (INTERPAS), nv, Buitenhaven 25, s-Hertogenbosch, The Netherlands) monthly. Available in seven sections covering I: Agriculture, fishing, food; II: Mining, metals, chemistry; III: Textiles; IV: Prime movers, electric power, tools; V: Transportation, civil engineering, architecture, hygiene; VI: Communication, photography, measuring; VII: Office articles, printing, general merchandise. Each section, which consists of a number of data sheets, gives full filing details and an English translation of the title of each specification listed.

Journal of the Patent Office Technical Society (Patent Office Technical Society, 214 Lower Circular Road, Calcutta, India) irregular. Classified listing of patent applications of Indian origin.

La propriété industrielle nucléaire: atomic patent abstracts (Service Central de Documentation du Commissariat a l'Energie Atomique, 44 Avenue du President Kennedy, Paris 16ᵉ) two per month. Abstracts of specifications from twenty countries.

List of publications and patents (Western Utilization and Development Division, Agricultural Research Service, US Department of Agriculture, 800 Buchanan Street, Albany 10, California, USA) 1955, semi-annually. Agricultural reports and patents.

Mineral exploration, mining and processing patents (Oliver North, 812 South Ode Street, Arlington, Virginia, USA) annual. Comprehensive coverage of US, Canadian and British specifications and selective coverage of German, French, Japanese and Russian specifications.

National catalog of patents (Rowman and Littlefield, 84 Fifth Avenue, New York, NY, 10011, USA, in association with the Interdex Corporation). This is an exhaustive catalogue of US patent specifications, adequately described by its subtitle, ' a guide to the . . . patents granted during the period as described in the *Official gazette* grouped in the classes and sub-classes of the *Manual of classification* of the US Patent Office '. Within each class patents are given in the form of one major claim with a drawing. The *Catalog* will be published in the following groups : 1 chemical; 2 electrical; 3 mechanical; 4 transportation; 5 instruments of precision. The chemical, electrical and mechanical volumes for 1961 to 1963 have been published, and if the project is supported adequately, work will commence on the volumes to cover 1941 to 1960. The *National catalog* will facilitate the searching of US patent literature by bringing together in an organised form the massive body of information which is scattered throughout more than 5,000 copies of the *Official gazette*. Each volume of the *Catalog* will contain its own manual of classification, subject index, index of cross references, and specification number index.

The international index of patents (Rowman and Littlefield in association with the Interdex Corporation). In this index each patent number is listed under the class and sub-class number and the full descriptive heading by which it is classified by the United States Patent Office. Its scope in each field covers the period from the granting of the first patent in the field up to the end of 1960. The complete index of US patents will consist of approximately fifty volumes and an index of foreign patents will comprise approximately thirty volumes. The foreign patents will be grouped according to country but listed according to the US *Manual of classification*. It is hoped to publish the complete body of patents covering all fields of technology. The volumes covering the following classes have been published : chemical : United States, 1790-1960; chemical : foreign; electrical : United States, 1790-1960; electrical : foreign.

The chemical : foreign group in five volumes covers over 350,000 chemical patent specifications from forty seven countries. The *Index* will be of great value in its own right but is also intended to act as a substitute for the above *Catalog* until the full range of *Catalog* volumes are available. Both projects, which have been undertaken with the full co-operation of the USPO, will make the wealth of information contained in patent literature more easily available to the searcher.

Paint and resin patents: a monthly digest of patents relating to surface coatings and the raw materials used in them (R H Chandler, 42 Grays Inn Road, London WC1). Abstracts of specifications from ten countries.

Patent abstract series (US Department of Commerce, Washington 25, DC, USA). Original series in seven volumes covering US patent specifications to 1953, supplements issued each year. Abstracts of government owned patents which are available on licence to firms, royalty free.

Patent licensing gazette (Techni Research Associates Inc, Professional Center Building, Willow Grove, Pennsylvania, USA) six times per year. Notifications of US patents available for licensing.

Patentschau Faeserstoffe und Textiletechnik, Ausg A-C (Institut fur Faeserstoffe-forschung der DAW, Abt Wissenschaftlisches Berichtswesen, Kanststrasse 55, Teltow-Seehof, Germany). European, US and USSR patents on dyestuffs and textiles.

Product licensing index, formerly *Lapis* (14 Homewell, Havant, Hampshire) monthly. Lists British and foreign products and processes covered by patents which are available for licensing or sale.

Science citation index (Institut for Scientific Information, 325 Chestnut Street, Philadelphia, Pa, 19106, USA) quarterly. Lists all US patent specifications which have been cited by other patents or by the 1,200 journals processed by the *Index.* By using this tool it is possible to determine how the information disclosed in a particular US patent specification has been used.

SOCMA *Patent list* (Synthetic Organic Chemical Manufacturing Association, 261 Madison Avenue, New York 16, NY, USA) 1925-, weekly.

Tecnica e invencion (La Direccion General de Prensa, Princesa 14, Madrid 8, Spain) monthly. Classified list of abstracts of Spanish patents and utility models.

Uniterm index to US chemical and chemically related patents (IFI/Plenum Data Corporation, 1000 Connecticut Avenue NW, Washington, DC 20036, USA) 1950-. The index contains two volumes for each year. One volume is a double dictionary of

61

3

term pages which is updated bi-monthly and is cumulative. The companion volume contains the reprint from the *Official gazette* of the US Patent Office for ready reference. In addition to the book form the information is also available on magnetic tape for machine searching.

Most abstracting services include patent specifications within their coverage although due to the delays often experienced with such publications their use in patent information work is usually more suited to retrospective search rather than to current awareness. *Chemical abstracts* (*CA*) is particularly impressive in its treatment of patent specifications : each issue contains a numerical patent index, covering the publications of twenty five countries, and a patent concordance which is useful in tracing equivalents. Each time *CA* receives a specification that corresponds to another specification for which an abstract has already been published in *CA* it will not be abstracted again but included in the concordance. A useful checklist of over one thousand abstracting services is given in *Ulrich's international periodicals directory,* 13th ed 1969/70, Bowker.

The following is a selection of British and American scientific and technical journals which regularly include either abstracts or notifications of recently published patent specifications :

British chemical engineering
British engineer
Chemical age
Engineering and mining journal
Food technology
Gas journal
Glass
Glass industry
Glass technology
Industrial diamond review
Industrial lubrication
Journal of the society of dyers and colourists
Light metal age
Metal finishing
Mining magazine
Modern plastics
Paint manufacture

Paper maker
Plastics
Plating
Platinum metals review
Refrigeration and air conditioning
SPE journal
Soap and chemical specialties
Textile manufacturer
Tooling
Ultrasonics
Underwater journal and information bulletin
Vacuum
Welding and metal fabrication
Wire industry
Wire world international

Chapter six: Standards and specifications—evolution and nature

Although standards and specifications play an increasingly important role in modern industry, standardisation has always been an essential feature of human activity. Communication is only possible as a result of standardisation. Speech was probably the first standard of civilised existence, for without standardised speech, where the same vocal sound always carries the same meaning, it would be impossible for a man to understand his neighbour. When the written word became the more sophisticated method of communication, a standard alphabet was needed to enable man to record his verbal communications. The development of commerce between primitive peoples established standards of weight, measurement and currency. Early units of measurement were rather crude; the earliest known unit, the Egyptian cubit, was the distance from the average man's middle fingertip to his elbow, and the inch was originally the length of three grains of barley laid end to end. The first attempt at scientific standardisation of measurement was made in 1558, when the length of a certain bronze bar was decreed to be the British standard yard. The metric system of measurement and weight was established in France between 1790 and 1799. At first the old arbitrary system of measurement died hard. There was much popular resistance to the introduction of the metric system, and in 1840 the French government deemed it necessary to pass a law forbidding the use of any other system.

One of the earliest attempts at the standardisation of products in the English speaking world was the publication of the first edition of the *British pharmacopâeia* in 1613, a work which endeavoured to determine the optimum compositions of the drugs and chemicals used in medicine. Although some of the early medicines were composed of highly questionable ingredients, the BP was developed on sound chemical principles and has since been published in numerous editions; the present Food and

Drugs Act demands that chemicals manufactured for use in food and drugs in the United Kingdom must comply with the standards of the BP.

Eli Whitney was probably the first to develop commercially the idea of standardised interchangeable industrial parts, which he used in the manufacture of a consignment of 10,000 rifles produced for the US army in fulfilment of his first government contract in 1798. Each part, copied from the individual parts of a model musket, was interchangeable with its standardised counterpart in the other muskets. Joseph Whitworth was the first British engineer to advocate standardisation to achieve interchangeability of threaded parts in the engineering industry. In a paper read to the Institution of Civil Engineers in 1841 he urged British industry to accept a uniform system of screw threads in place of the various shapes and sizes of thread then obtaining. The unified system of screw threads, an adaptation of Whitworth's thread, is now in use throughout Great Britain, the USA and Canada. Sir John Barry, a civil engineer, convinced of the necessity to standardise structural components, persuaded his industry to 'consider the advisability of standardising various iron and steel sections', and thus was formed the Engineering Standards Committee which was the precursor of the British Standards Institution (BSI).

STANDARDS AND SPECIFICATIONS

Today many thousands of accepted standards are widely used in science and technology, and standards and specifications now occupy an essential place in technical literature. Standards and specifications are documents which state how materials and products should be manufactured, defined, measured or tested; they are documents which lay down sets of conditions which should be fulfilled. Standards have emerged whenever repetitive operations have been developed as organised procedures.

A specification has been defined by the American Standards Association (ASA), now the American National Standards Institute (ANSI), as 'a concise statement of the requirements for a material, process, method, procedure or service, including, whenever possible, the exact procedure by which it can be determined that the conditions are met within the tolerances specified in the statement; a specification does not have to cover specifically

65

recurring subjects or objects of wide use, or even existing objects '. A standard is an evolution of a specification and is defined by the same body as ' a specification accepted by recognised authority as the most practical and appropriate current solution of a recurring problem '.[1] The standard may only remain effective for a limited period, because it may become obsolete due to progress in technology; better methods of testing or improved techniques may be developed, more suitable materials may be discovered for a specific purpose. Standards are essentially dynamic, not static, and if a better method of carrying out an operation or testing a material is found it will be codified into a new standard.

MASS PRODUCTION AND STANDARDISATION

Mass production would have been impossible without standardisation. The prosperity of the American economy dates from the time when men like Henry Ford realised the immense market potentialities of mass produced standardised goods. Standardisation avoids wastage of resources and there is no doubt that time and money are saved when a standard is available offering a common language, a judicious procedure and interchangeability or a guarantee of fitness for use.

In earlier days, a craftsman worked as an individual, producing objects which sought to achieve perfection as a whole through the balance of unspecified parts. Modern industry works as a team, each man producing quantities of a single standardised component which can be assembled into a complete finished product. The availability of standard mechanical parts—slabs, plates, bars, etc—all manufactured according to metallurgical and dimensional standards, enables a designer to create new end products, rather in the manner in which a child uses a meccano set. Standards can release for new creative work manpower which might otherwise have been used to reinvent standards. An official of the Ford Motor Company has stated, exactly, that his firm does not make standard cars, it manufactures standard parts.

Many case histories of increased productivity through standardisation have been reported, savings having been mainly achieved through ' variety reduction '. South Wales Switchgear Ltd recently applied standardisation to over ninety percent of their manufactured components, so that nearly every part manufactured in the entire voltage range 3·3 to 11 kv is now inter-changeable; in three years the firm increased its productivity by fifty percent. Standardisation also avoids waste of any form; the recent introduction of standardisation by a glass firm resulted in the amount of their bottle scrap being reduced from sixteen percent to two percent. A survey of company standardisation carried out in 1956 by ASA estimated that a company would save $6 for every other dollar spent on the standardisation programme.

TYPES OF STANDARD AND SPECIFICATION

Standards and specifications may be conveniently divided into : dimensional standards; material standards; standards of perfor-mance or quality; standards of testing; standards of nomenclature or terminology; codes of practice.

1 *Dimensional standards* specify the dimensions needed to achieve interchangeability or to make things fit, *eg* nuts and bolts; dimensional standards for machine components facilitate the easy replacement of used or damaged parts.

2 *Material standards* cover the chemical composition, condi-tion, tolerances and mechanical properties of raw materials such as alloys, steels and pigments.

3 *Standards of performance or quality* will ensure that a product is adequate for its intended purpose, *eg* that domestic paraffin heaters conform to guarding and safety requirements, or that switchgear equipment is manufactured according to mini-mum insulation standards.

4 *Standards of testing* enable materials and products intended for the same purpose to be compared; methods have been formu-lated, for example, for the testing of the elasticity, viscosity, shrinkage, tensile strength etc of materials.

5 *Standardised terminology* enables scientists and technolo-gists working within a certain industry to communicate more exactly. A standard vocabulary will ensure that a term has a unique, unambiguous meaning. Symbols used in engineering must be standardised if confusion is to be avoided, and therefore several standard compilations of symbols have been issued by national and industrial organisations.

6 *Codes of practice* cover the installation and maintenance of equipment, or cover the correct method of accomplishing a certain

task; thus codes of practice exist for such divergent subjects as the protection of buildings against lightning, or the laying of tiles.

DOCUMENTATION STANDARDS
In recent years there has been a quickening of activity in the field of documentation standards. Librarians and information workers have realised that standardisation is indispensable if cooperation is to be effected in the development of information services. They have also become aware of the fact that the application of standards at source can prevent the escalation of documentation problems. In the UK the Office of Scientific and Technical Information (OSTI) has supported BSI's documentation programme, while in the United States the Committee on Scientific and Technical Information (COSATI) has set up a subpanel on standardisation to review the role of COSATI in the development of documentation standards for implementation throughout the Federal government's information services. Documentation standards can be classified into the following categories: i) microphotography and documentary reproduction, *eg* legibility of microcopies; ii) editorial and bibliographical methods, *eg* proof correction, bibliographical references, abbreviations of titles of periodicals, etc; iii) data processing, *eg* alphanumerical punching codes for data processing cards; iv) transliteration; v) classification and cataloguing codes, *eg* the Universal Decimal Classification.

REFERENCE
[1] *Standardization* (41) January 1960 10.

Chapter seven: Standardising bodies

Standards and specifications are issued by: companies, industries, trade associations and technical societies; government departments; national standardising bodies; international standardising bodies.

COMPANIES

Company standards are issued by the standards department, which is ' a company department charged with the duties of defining the parameters of standardisation within the company, preparing for management approval a standardisation programme and policy, and implementing the programme within the company's jurisdiction '.[1] The standards department will implement the use of a national standard within the company and also develop the company's own standards when no suitable national standard exists which covers the needs of the firm. The standards department may need to modify a national standard to meet its own peculiar needs, and in such cases it may issue a company document which is simply an edited version of the national standard, with a company reference number assigned. The company standard reference number will act as a technical shorthand and will obviate the need to repeat details of intricate procedures or dimensions.

The standards department will be supervised by the *standards engineer*. He will be responsible for promoting, developing and disseminating information about company standards, for screening all national and trade standards and also for demonstrating to management the advantages of a particular standardisation project. As standardisation covers the whole field of technology and demands specialised knowledge of a variety of fields which no single standards engineer can hope to acquire, the latter will co-operate with subject experts to produce draft standards and specifications and then consult production engineers to ensure that the drafts are acceptable. A gratis BSI booklet entitled *Operation of*

69

3*

a company standards department, which includes sections on the function of the department, the preparation of standards and variety control, was published in 1970.

The British Standards Institution invites all standards engineers to become members of the Standards Associates Section which exists to: 1 provide a forum for the exchange of views on standards matters; 2 ensure closer co-operation between BSI and those using standards; 3 promote the wider implementation of standards techniques; 4 disseminate information on new developments in standards.

INDUSTRIES, TRADE ASSOCIATIONS, TECHNICAL SOCIETIES
Standards are issued by trade associations and technical societies to meet the needs of the industries which they serve. The Society of Motor Manufacturers and Traders (SMMT), established in 1955, reviews the standards used by the British motor industry. Some of the standards adopted are SMMT's own, while others are national standards issued by BSI or standards published by bodies such as the American Society of Automotive Engineers. Other examples of specifications produced for a specific industry are the British Road Tar Association's standards which include a specification for ' Dense tar surfacing of roads ' and the Cement and Concrete Association's specification for ' Housing estates and other minor roads in concrete '.

In the USA the Society of Automotive Engineers (SAE) claims that its standards programme has been responsible for enhancing vehicle safety, simplifying maintenance and reducing production costs. The Trade Association of the Pulp and Paper Industry (TAPPI) issues a volume of standards which cover testing methods recommended for use within the industry. The National Electrical Manufacturers Association (NEMA) standards cover the construction, testing, performance and manufacture of ac and dc electric motors and generators.

If specifications and standards contain requirements, methods of testing these requirements must be available; the American Society for Testing and Materials (ASTM), founded in 1898, is an international and non-profitmaking technical, scientific and educational society devoted to ' the promotion of knowledge of the materials of engineering and the standardisation of specifications and methods of testing '. A collection of ASTM standards known as the *Book of ASTM standards* is issued annually, the current edition (1971) being in thirty three volumes and containing 4,300 standards and tentatives. The standards relating to a specific field are collected together in each volume; thus part one contains standards on steel piping materials; part two, ferrous castings. The standards in each volume are arranged in numerical order of standard number and each volume is equipped with its own subject index. A separate index to the complete range of ASTM standards is issued annually.

An industrial or technical society standard is often adopted as a national standard. For example, SMMT standards may be published later as British standards after they have been submitted to BSI for approval and if they are considered as suitable for use in other industries. A list of SMMT standards which are being considered for approval as British standards appears regularly in BSI *News*. Many standards issued by American technical societies are subsequently reissued under the imprint of ANSI, *eg* American national standard P2.2-1964 was originally issued as ASTM D 586-63. A particularly useful set of German technical society publications are the *VDI Richtlinen (German engineering guidelines)* issued by Verein Deutscher Ingenieur, the German Association of Engineers. The scope of these guide lines is similar to that of a British Standards code of practice, but they include more detail, embracing, for instance, instructions to technical personnel on plant operation. They differ from German national standards (DIN) in that they reflect the current state of the art of a particular topic and are not written in the form of recommendations. An English language index to the Richtlinen is available from the Department of Trade and Industry, Abell House, John Islip St, London SW1. Subjects covered by the guide lines include production engineering, power engineering, heating, design, noise abatement, metrology and vehicle technology.

GOVERNMENT DEPARTMENTS

Government departments issue standards covering the field of their jurisdiction. S K Reeves asserts that as far as the armed forces are concerned the purpose of issuing government specifications and standards is ' to secure the procurement of standardised stores of the requisite quality and the application of standardised technical procedures by the Services, with a view to achieving

71

economy and efficiency over a very wide field '. The Ministry of Aviation publishes DTD *Aerospace material specifications*. The specifications are issued 'to meet a limited requirement not covered by an existing British standard or to serve as a basis for inspection of material, the properties and uses of which are not sufficiently developed to warrant submission to BSI for standardisation '. When a DTD specification becomes stable and its usage is widespread it is offered to BSI for standardisation.

The most important series of British government standards are the *Defence standards* now being issued by the Directorate of Standardization of the Ministry of Defence to supersede most of the heterogeneous series of specifications, lists, guides and codes of practice formerly issued by the various service departments of the Ministry of Defence and the Ministry of Technology. Reeves' previously cited article is an excellent survey of this maze of publications and his appendix contains a useful list of sources of supply of the listed documents. The *Defence standards* are the means of 'promulgating the Services' requirements for design and performance of equipment or components; for specifying standard ranges of items; and for establishing codes of practice. They are also the means of implementing international military standardization agreements '.[2]

The US Defense Standardization Act 1952 aimed at standardising effort throughout the US military services. At present over 20,000 *Military standards and specifications* are in use. Although the bulk of the Department's activity is directed towards military science and combat items many *Military specifications and standards* are directly applicable to civilian industry and consequently these documents are regularly used by commercial firms. The military does not use its own specifications exclusively; whenever a suitable industry standard or specification exists it is Department of Defense policy to use it. In certain circumstances, however, special military standards are essential, *eg* when industrial standards specify a minimum quality which would not fare well under rough military use. The US Bureau of Ships was concerned by the duplication of some industry and military standards and placed a contract with ASA to develop a cross-index of military and related industry standards. The index appeared as a monthly feature in *Magazine of standards* during the nineteen sixties.

The General Services Administration, Federal Supply Service, Standardization Division issues *Federal specification and standards* to cover all materials and supplies required by civil departments of the government and also by military agencies where needed. *Federal standards* are issued within five groups: supply standards (qualities, types, sizes, etc); test method standards; engineering standards; procedural standards. Again many *Federal standards* are applicable in private industrial situations.

NATIONAL BODIES

Most countries have established national standardising bodies which are responsible for promoting national standards for application within the country. Before the first world war, only one body, the Engineering Standards Committee, the forerunner of BSI, existed. By 1918 two other bodies, the Deutscher Normenauschuss (DNA) and the American Engineering Standards Committee (ASA) had been established. Today national standardising bodies exist in more than fifty countries.

BRITISH STANDARDS INSTITUTION (BSI)

BSI is the national standardising organisation for the United Kingdom. It was founded as the Engineering Standards Committee by the Institution of Civil Engineers in 1901. As the need for standardisation became accepted, the committee extended its activities beyond the bounds of civil engineering into the mechanical and electrical fields. A royal charter was granted in 1929 and the title British Standards Institution was adopted in 1931. Today BSI's range of activities cover every major field of industry. BSI's functions as defined in the charter are:

1 To co-ordinate the efforts of producers and users for the improvement, standardisation and simplification of engineering and industrial materials, so as to simplify production and distribution, and to eliminate the national waste of time and material involved in the production of an unnecessary variety of patterns and sizes of articles for one and the same purpose.

2 To set up standards of quality and dimensions, and prepare and promote the general adoption of British standards, specifications and schedules . . . and from time to time to revise and amend such specifications and schedules as experience and circumstance may require.

73

3 To register, in the name of the Institution, marks of all descriptions, and to prove and affix or license the affixing of such marks or other proof, letter, name description or device.

In short, BSI's main function is to formulate standards and codes of practice which may be voluntarily adopted by interested organisations. The standards are prepared by 4,200 technical committees, their 22,000 members being drawn from industry, research institutions, consumer groups etc. As noted above, the technical committees' function is not completed once the standard is published, for they continue to be responsible for reviewing relevant existing standards which may have been overtaken by technological progress. Each technical committee is responsible to the appropriate industry standards committee (ISC) which initiates and approves a standards project. The ISC are grouped under divisional councils: building, chemical, engineering, textile, miscellaneous (ie consumer goods, personal safety), codes of practice.

Requests for the establishment of a British standard come from many sources—manufacturers' associations, government departments, consumer groups, users, or an established BSI technical committee. BSI does however have two priorities in its choice of projects: i) the development of metric standards, ii) the formulation of harmonious British standards to cover areas in which international recommendations have been published by the International Organisation for Standardisation or a similar body.

The preparation of the standard can involve the following steps:

1 Request from responsible authority;
2 ISC approves request and sets up a technical committee;
3 Technical committee carries out the initial investigation and research;
4 Draft standard is prepared;
5 Draft standard circulated to interested parties for comment;
6 Standard amended in the light of comments received;
7 Amended standard sent to ISC for approval;
8 New standard published.

A British standard has no mandatory implications; it is not an instrument for enforcing government control. BSI is an independent body and its publications are simply a basis for voluntary application.

BSI is financed by:

1 Income from its 15,000 subscribing members (membership at £12 per annum is the obvious way for an organisation to keep itself informed of standards activity in the United Kingdom);

2 A government grant;

3 Sales of publications;

4 Certification mark and testing fees.

The total income for 1968/69 was £2,070,000, of which £290,000 was accounted for by fees for certification and other services.

There are almost 7,000 British standards currently available and an average of 500 new standards are published annually. Some 2,000 standard projects are currently under consideration by technical committees.

The presence of the BSI ' *Kite Mark* ' on a manufactured object will assure the purchaser that the product has been manufactured in accordance with BSI requirements. The products are certified by independent inspection; the granting of a Kite Mark to a firm does not imply once and for all approval, as regular inspection is undertaken by BSI to ensure that quality is maintained. The Kite Mark is valuable to manufacturers and purchasers, alike. BSI claims that ' it reinforces the manufacturers' own quality control and provides an independent assurance of compliance with a standard; it obviates the need for the individual sampling and testing by industrial purchasers, local authorities, government departments, etc '. The mark is now in use in association with over 260 British standards. Although British standards have no mandatory implications *per se,* government legislation is passed from time to time which makes the marking of certain products compulsory, examples of this being car safety belts, with BS 3254 and motor cycle helmets with BS 2001.

The testing for compliance with British standard requirements is undertaken at the Hemel Hempstead Centre. The government-supported Technical Help to Exporters (THE) service has been administered from the same location since 1966. THE is staffed by a team of experienced engineers and exists to provide manufacturers with advice and up-to-date information on overseas specifications and regulations, particularly where an approval scheme is in operation. THE's areas of operation include:

75

i) The *Technical digest service* which consists of a series of information sheets each covering a subject such as Boilers and pressure vessels, Cranes, Gas equipment, etc, summarising the specifications, regulations and approval systems existing in a particular country;

ii) The translation of foreign standards and regulations;

iii) Assistance in obtaining test certificates or approval for products overseas;

iv) The identification, provision and assistance in the interpretation of foreign standards and technical requirements;

v) Assistance with specific technical matters relating to export problems;

vi) The arrangement for the testing and inspection in the UK of products for export.

AMERICAN NATIONAL STANDARDS INSTITUTE (ANSI)

The precursor of ANSI was the American Engineering Standards Committee founded in 1918 by the American Institute of Electrical Engineers, the American Society of Mechanical Engineers, the American Society of Civil Engineers, the American Society of Mining and Metallurgical Engineers, and the American Society for Testing and Materials, to co-ordinate the development of national standards. The name American Standards Association (ASA) was adopted in 1928 and the title was later changed to the United States of America Standards Institute (USASI). The present title was adopted on the 6th October 1969 as it was thought that the designation United States of America Standards Institute might imply that the organisation was a government agency. This possibility would be completely incompatible with ANSI's projected certification marking programme. Under the programme ANSI will license manufacturers to use its mark on products which have been independently tested and been found to comply with American national standards. Only those products which conform to standards which include safety specifications will be accepted for certification. ANSI is now an association of technical societies, trade associations and company members. Fourteen standards boards, *eg* acoustical, chemical etc, review and consider all matters concerned with standards that fall within their province, and their findings and recommendations are submitted to the Standards Council for approval.

76

The functions of ANSI are defined as:

1 To provide systematic means for developing American standards;

2 To promote the development and use of national standards in the USA;

3 To co-ordinate all national standardising activities;

4 To serve as a clearinghouse for information on American and foreign standards;

5 To represent US interests in international standardisation activity;

As with British standards, all American national standards are arrived at by common consent and are available for voluntary use. An American standard can be established by one of three procedures:

1 *The sectional committee method*: a committee consisting of representatives of all interested organisations formulates a draft standard. When criticism has been considered and the draft modified, the committee votes by ballot on the final draft and, if a consensus of opinion is obtained, the draft is submitted to the Standards Council for approval.

2 *The existing standard method*: an existing standard of any technical society can be submitted to the Standards Council for approval as an American standard.

3 *The general acceptance method*: intended for uncomplicated standards which can be examined at a general conference and approved without prolonged discussion.

More than 4,000 American standards are currently available with 400 standards projects under consideration.

PHYSICAL CONSTANTS

In addition to national standardising bodies which formulate standards for industrial and general application, many countries have established organisations to formulate and maintain standards for the determination of physical constants. In the USA the Institute of Basic Standards of the National Bureau of Standards (NBS) is responsible for maintaining the national standards of physical measurements upon which all measurements used in the USA are based and the determination of physical constants and properties of materials. The institute is organised into thirteen scientific divisions, each corresponding approximately to a major field of physical science or engineering, and these sections are again divided into sub-sections, each responsible for a technical area within each field. NBS co-operates with recognised standardising bodies such as ANSI and ASTM. NBS research provided the foundations for the *National electrical safety code* (an American national standard) and other codes of national or local application. Broadly, the NBS deals with scientific standards while standards of style, convenience and policy are the province of other organisations.

The results of the bureaux' work are published as:

1 Papers published in NBS journals, *eg Journal of research;* such papers are also reprinted as RPS—*Research papers.*

2 Circulars, *eg* C 466 *American standard specification for dry cells and batteries.*

3 Handbooks: recommended codes of engineering and industrial practice, *eg* H 46 *Code for protection against lightning.*

4 Building materials and structural reports, *eg* BM 3118 *Ventilation of plumbing fixtures.*

5 Applied mathematics series: usually mathematical tables, *eg* AMS 8 *Tables of powers for complex numbers.*

The publications of NBS are catalogued in *Circular 460: publications of the NBS 1901 to June 30 1957* and in *Miscellaneous publication 240: publications of the NBS July 1 1957 to June 30 1960.* This is regularly brought up to date by supplements.

NATIONAL PHYSICAL LABORATORY

The Measurement Group of the National Physical Laboratory performs similar functions for the United Kingdom as the NBS performs for the United States. The division exists ' to provide standards for a wide variety of quantities in mechanics, electrostatics, temperature etc, also to determine various physical quantities and constants '. The division also co-operates with BSI in the regular preparation of draft standards for engineers' measuring equipment and similar apparatus. The results of the division's work are published either as papers in a wide range of scientific journals, or as *NPL Notes on applied science, eg* 24: *Photometric standards and the unit of light.* Physical constants are published in four booklets as *Units and standards of measurement employed at NPL:* 1 ' *Mechanics—length, mass, time and fre-*

78

quency, angle, volume, density . . .'; 2 ' Light '; 3 ' Electricity ';
4 ' Temperature '.

A complete listing of NPL publications is given in HMSO *Sectional list no. 3: Department of Trade and Industry.*

INTERNATIONAL ORGANISATIONS
National standards can be a barrier to the interflow of trade between countries, for although the differences in national standards for the same article may only be minute, they render interchangeability impossible, thus making the marketing of the same product in a large number of countries also impossible. A senior official of the Phillips group of companies has illustrated this by pointing out in a recent publication that different national standard requirements for radio receivers necessitate the manufacture of separate versions of the same model to suit the requirements of each market.[3]

International Electrotechnical Commission (IEC): Electrical engineering was the first technical profession to concern itself with international standardisation and the IEC was founded in 1906 to facilitate the co-ordination and unification of national standards. IEC publishes recommendations which express the international consensus of opinion and which are intended to serve as the basis for the national standards of the member countries. The work of IEC covers the whole field of electrotechnology, extending into fields of electronics, telecommunications and nuclear energy. The commission is now composed of thirty eight national committees.

International Commission on Rules for the Approval of Electrical Equipment (CEE): CEE, whose membership is at present fifteen countries (limited to European nations), was founded in 1926 and is concerned with the conditions with which certain types of electrical equipment should comply in order to protect the public against hazards resulting from equipment of poor quality. An agreement has been ratified by IEC and CEE which co-ordinates the work of both bodies and obviates duplication of effort, and this co-ordination has resulted in the publication of some specifications which are common to both bodies.

The European Electrical Standards Coordinating Committee (CENEL) was formed in 1960 with its membership recruited from the various national committees concerned with standardisation in electro-technology. The work of CENEL complements that of bodies such as IEC and CEE: a preliminary examination is made of each relevant recommendation published by the international bodies. If significant differences are still found to exist between national practices and the recommendation a questionnaire is circulated to ascertain the degree of a conformity which exists in each country. After the survey has been completed the replies are analysed in an effort to clear the ground for reaching complete agreement between CENEL member countries. The originating body is then approached through its own national committee member in an attempt to bring about any changes in the recommendation that would result in complete agreement.

International Organisation for Standardisation (ISO): ISO was established in 1946 ' to reach international agreement on industrial and commercial standards and thus facilitate international trade as well as the interchange of scientific and technological data relevant to standards '. Membership consists of fifty two national standards organisations, and ISO's work is undertaken by more than 100 technical committees (TCs), each covering a special field, *eg* acoustics, documentation, terminology. The members of the TCs, who are drawn from the national bodies, formulate draft recommendations which are studied by each national body before publication. To date some 450 recommendations have been published. ISO recommendations are often rather brief and are intended to act simply as a framework upon which a more elaborate national standard may be constructed. CERTICO was ' established in 1970 as a special committee of ISO in an attempt to implement the mutual acceptance by European countries of the validity of national certification marks'. CERTICO's programme, which plans harmonious procedures for certification and approval schemes, should have a salutary effect on the expansion of international trade.

In 1960 the countries of the European Economic Community (EEC) and the European Free Trade Association (EFTA) combined to form the European Standards Co-ordinating Committee (CEN). The purpose of this organisation is to provide the machinery for harmonising national standards in sufficient detail to remove technical barriers to trade. Progress is being made towards a European standard which would be accepted by each of the CEN countries as a national standard. It is not intended that the work of CEN should conflict with the efforts of ISO, to which all the member countries of CEN belong. ISO recommendations will be automatically adopted and thus it is claimed that CEN would encourage the implementation of international agreements in a highly developed economic region.

The USA, Britain, Canada and Australia have co-operated in engineering fields to produce mutually acceptable standards. These specifications, the best known of which is the ' unified ' screw thread, are known as ABCA standards. The need for such standards became apparent during the second world war, when it was realised that American military equipment could not be serviced with British spares. The NATO countries have also arrived at several agreed standards which have been published as STAG-NAGS. ABCA and STAGNAGS for application in the United Kingdom are issued in the SDM (' Standardisation design memoranda ') and SSM (' Standard stores memoranda ') series, published by the Ministry of Defence. SDM and SSM specifications are being slowly superseded by Defence standards but many of them will remain in force for some time until the superseding standard has been issued. Lists of newly published Defence standards appear in the monthly *Standardization,* available from the Ministry of Defence, Directorate of Standardization.

Other examples of international standardising bodies working in more specific subject fields are the International Bureau for the Standardisation of Man-Made Fibres, which publishes methods of testing and rules for the classification of man-made fibres, and the International Conference for Promoting Technical Uniformity on Railways, which issues its annual technical standards.

REFERENCES
[1] *Standardization* (41) January 1960 11.
[2] Reeves, S K: Specifications, standards and allied publications for UK military aircraft. *Aslib proceedings* 22(9) September 1970 432-48.
[3] Woodward, C D *editor*: *Standards for industry* (Heinemann, 1965) 16.

Chapter eight: Bibliographical control of standards

Standards are invariably referred to by a prefix and number which is assigned to the standard by its originator. The main problems encountered by librarians and information officers when they are endeavouring to obtain standards for their readers are those of identification and location, *ie* of decoding the number by which the standard is requested so that they can make sure that the number quoted does not refer to a technical report, and then if it has been established that the document is a standard, to ascertain its originator and to find a location from which a copy of the standard can be obtained. A prospective standards user will not usually ask whether a library has the American national standard on seamless copper tubes for refrigeration field service, the international specification for electric cooking and heating appliances for domestic purposes, or the Ministry of Defence specification on valve holders and wiring jigs, but simply for ANSI H23.5-1963, CEE II, or DEF-5251.

NATIONAL STANDARDS

A British standard might be cited as, for example, BS 2509 : 1960 and an American national standard as ANSI C8.1-1944. For the British standard, the prefix BS obviously refers to the issuing body; 2509 is the running number which is allocated to British standards, the main sequence being numbered from 1 onwards; 1960 indicates the year the standard was approved. Other prefixes used with British standards are :

AU—	Automobile series;
A/X—(*eg* A29; G171)	Aerospace series;
BS/MOE-	Standards prepared for the Ministry of Education in collaboration with Ministry of Works;
CP—	Codes of practice;
MA—	Marine series.

For the American national standard, ANSI denotes the issuing body; c8.1 the standard number. American national standards are classified into nineteen groups and allocated a number within the group, and c is the electrical engineering group, c8 the number for insulated wire. Within the number ANSI c8 the standards are issued numerically as they are approved and published, eg C8.1; C8.9; C8.12; -1944 indicates the year of approval.

Standards issued by other national bodies are codified by the following prefixes:

ABNT   *Brazil*: Associacoa Brasileira de Normas Technicas, Caixa Postal No 1680, Rio de Janeiro.

AS   *Australia*: Standards Association of Australia, Science House, Gloucester and Essex Streets, Sydney.

BDS   *Bulgaria*: Comité Superieur de Normalisation de la Republique Populaire de Bulgarie, 5 Rue Idanov, Sofia.

CSA   *Canada*: Canadian Standards Association, 235 Montreal Road, Ottawa 2.

CSN   *Czechoslovakia*: Urad pro Normalisaci, Vaclavske Namesti c 19, Prague 3.

DGN   *Mexico*: Direccion General de Normas, Av Cuauhtemoc No 80, Mexico 7.

DIN   *Germany* (FDR): Deutscher Normenausschuss, Uhlandstrasse 175, Berlin W15.

DS   *Denmark*: Dansk Standardiseringsraad, Vesterbrogade I, Copenhagen.

ENO   *Greece*: Comité Hellenique de Normalisation, Aupres de la Chambre Technique de Grece, Rue Kolokotroni 4, Athens.

EOS   *Egypt*: Egyptian Organisation for Standardisation, 144 Tahrir Street, Dokki, Cairo.

GOST   *USSR*: Komitet Standartov, Mer i Izmeritel' nyh Priborov Pri Sovete Ministrov SSSR, Leninskij Prospekt No 9B, Moscow V 49.

INDITECNOR *Chile*: Instituto Nacional de Investigaciones Tecnologicas y Normalizacion, Teatinos 20, 2 Piso, Santiago.

IRAM   *Argentina*: Instituto Argentino de Racionalizacion de Materiales, Chile No 1192, Buenos Aires.

IS	*India*: Indian Standards Institution, Manak Bhaven, 9 Mathura Road, New Delhi 1.
IS	*Ireland*: Institute for Industrial Research and Standards, Glasnevin House, Ballymun Road, Dublin.
JIS	*Japan*: Japanese Industrial Standards Committee, Agency of Industrial Science and Technology, 3-1 Kasumigaseki, Chiyoda-Ku, Tokyo.
JUS	*Yugoslavia*: Jugoslovenski Zavod za Standardisaciju, Post Pregradak 933, Belgrade.
MZS	*Hungary*: Office Hongrois de Normalisation, Ulloi-ut 25, Budapest IX.
NBN	*Belgium*: Institut Belge de Normalisation, 29 Avenue de la Bradanconne, Brussels 4.
NEN	*Netherlands*: Stichting Nederlands Normalisatie-Instituut, Duinweg 20/22, Postbus 70, The Hague.
NF	*France*: Association Française de Normalisation, 23 Rue Notre Dame des Victoires, Paris 2.
NI	*Indonesia*: Dewan Normalisasi Indonesia, Djalan Braga 38, Bandung.
NORVEN	*Venezuela*: Comision Venezolana de Normas Industriales, Direccion de Industrial, Ministerio de Fomento, Caracas.
NP	*Portugal*: Reparticao de Normalizacao, Avenida de Berna 1, Lisbon.
NS	*Norway*: Norges Standardiserings-Forbund, Haakon VII's gt 2, Oslo.
NZSS	*New Zealand*: New Zealand Standards Institute, PO Box 195, Wellington C1.
ONORM	*Austria*: Oesterreichischer Normenausschuss, Bauernmarkt 13, Vienna 1.
PN	*Poland*: Polski Komitet Normalizacyjny, UI Swietokrzyska 14, Warsaw 51.
PS	*Pakistan*: Pakistan Standards Institution, 39 Garden Road, Saddar, Karachi 3.
SABS	*South Africa*: South African Bureau of Standards, Private Bag 191, Pretoria.
SFS	*Finland*: Suomen Standardisoimisliitto RY, Ratakatu 9, Helsinki.

SI	*Israel*: Standards Institution of Israel, 200 Dizengoff Road, Tel Aviv.
SIS	*Sweden*: Sveriges Standardiseringskommission, Box 3295, Stockholm 3.
SNV	*Switzerland*: Association Suisse de Normalisation, General Willestrasse 4, Zurich 2.
STAS	*Rumania*: Oficiul de Stat Pentru Standarde, Str Joliot Curie 6, Bucharest 30.
TS	*Turkey*: Turk Standartlari Enstitusu, Adakale Sokak No 27/3, Yenisehir, Ankara.
UNCO	*Columbia*: Instituto de Normas Columbîana, Division de Investigaciones Cientificas, Ciudad Universitaira, Universtood Industrial de Santander, Bucaramanga.
UNE	*Spain*: Instituto Nacional de Racionalizacion del Trabajo, Serrano 150, Madrid 6.
UNI	*Italy*: Ente Nazionale Italiano de Unificazione, Piazza Armando Diaz 2, Milan.

NATIONAL BODIES : CATALOGUES

Most national standardising bodies publish an annual list of their standards in a yearbook or catalogue. BSI publishes its BS *Yearbook* which contains a numerical list of the various series of British standards; an abstract of each standard and recommendation is also given. A subject index to the list of British standards enables an enquirer to ascertain which standards have been published on a specific subject. The publications of international standardising organisations are listed and summarised in a separately published supplement which contains a table showing British standards and international recommendations on related subjects and an alphabetical subject index. It also contains a UDC index to both British standards and ISO recommendations.

BSI also issues *Sectional lists of standards* for a number of broad subjects, such as chemical engineering, drawing practice and plastics. *British standard handbooks* which collect together under a single cover either the complete text of related standards, or summaries of these standards, are published for certain subjects, *eg* building materials and components, methods of test for textiles, etc. British standards relating to broad technical fields can also usually be traced by referring to the relevant trade annual or directory, *eg Metal industry handbook and directory, British plastics yearbook, Soap, perfumery and cosmetics yearbook*. The value of these lists in tracing standards would be enhanced if they were extended to cover standards other than those issued by BSI, as in many fields the national standards represent only a fraction of those frequently requested by library and information department users.

The annual *Catalog of the American National Standards Institute* lists standards by their prefixes within the nineteen groups, and also covers ISO, IEC and CEE recommendations. Each issue of the *Catalog* is equipped with an alphabetical subject index. Deutsches Normenausschuss issues DIN *Normenblatt-Verzeichnis*, covering German standards, which classifies the standards by UDC then gives a brief title listing. The main classified sequence is supplemented by a standard number index and an alphabetical subject index. A difficulty which technical librarians experience is that when an enquirer requests a foreign standard by a prefix number only, he may not realise that the document he is asking for is in fact in a foreign language, and therefore when the specification is obtained it may be quite useless to him without a translation. DNA has anticipated this situation and has published many of its standards in English as well as German; an annual catalogue of *DIN English translations of German standards* is also published, arranged in the same manner as the *Normenblatt*. BSI have produced a cover-to-cover translation of the Russian standards catalogue. Although this has not been made generally available the master copy can be consulted in the library. The Association Française de Normalisation publishes its *Catalogue des normes Français*, a classified list with an alphabetical subject index.

Most national bodies also publish a periodical which supplements their yearbook by giving details of new standards, standards under preparation, etc. BSI *News* contains current information on standardisation, details of new BS publications and details of new work started. BSI *List of overseas standards* gives a classified sequence of new and draft standards received from abroad in the BSI library. ANSI issues a bi-monthly *Listing of new and revised American national standards* to supplement its *Catalog*. *DIN—mitteilungen: zentral organ der deutschen normung* in addition to including notifications of new German standards gives a UDC classified listing of overseas national and international publications.

In addition to the national sources mentioned above, the recommendations of international bodies may be traced through their catalogues and lists. ISO publishes annually its *Memento,* which gives details of member bodies, ISO technical committees, while the annual catalogue gives numerical and classified listings of ISO recommendations and draft recommendations. The catalogue is up-dated each month by *ISO bulletin.* IEC publishes an annual *Catalogue of publications* and a list of CEE specifications is issued periodically.

LOCATION
Once a national or international standard has been identified, it can be borrowed from the BSI library, which maintains a complete loan collection of all standards issued by overseas national standardising bodies in addition to sets of the most important series such as *ASTM standards* and *US federal specifications and standards.* The BSI library does not hold copies of non-official standards, but functions as a clearinghouse for information on these standards and will endeavour to put an enquirer in touch with the organisation which can supply a copy of the standard, thus an enquirer for *US military specifications* could be directed to the Department of Trade and Industry Technology Report Centre (TRC) at St Mary Cray, Kent. The library also maintains an alphabetical subject index of foreign national standards, from which it is possible to ascertain which standards have been published throughout the world on a specific subject. Complete sets of British standards are maintained in most large public technical libraries in the British Isles. The largest technical libraries will also carry sets of ANSI, DIN and ASTM standards. The ANSI library serves as a clearinghouse for information on standards from every source, in addition to maintaining a loan collection of over 100,000 standards.

The Committee for Index Cards (CICS) of the ISO have published *Rules for allocating UDC numbers to standards and for compiling catalogues of standards,* and *Rules governing standards cards,* to promote a systematic exchange of index cards between member bodies, and thus facilitate the use and co-ordination of standards issued by national bodies. At present nineteen national bodies participate in this exchange scheme, which gives each of the countries a central index for identifying national standards on related subjects.

The standards produced by government departments, technical societies and other bodies are more difficult to identify than national standards which are adequately documented. The most important bodies may issue catalogues of their standards, examples being ASTM's *Index to standards* and the *Index to defence specifications* published by the Ministry of Defence. This latter index is supplemented by *Standardization,* a periodical which gives details of new DEF standards. In many cases, however, the librarian must rely on his own resources to identify non-official standards. One possible short cut to identifying and locating a standard may be to contact the relevant technical society or trade association when the subject matter of the specification would interest them, *eg* the Institution of Heating and Ventilating Engineers for air conditioning specifications.

As many standards are available only from their issuing body and cannot be obtained from any library, there is a need for a comprehensive decoding guide which would list the standard prefixes, then indicate where copies of the standards could be obtained. The closest equivalent to this type of publication is the DIN *AGkürzungen: technisch-wissenschaftlicher organisationen des in- und auslandes und iker veroffentlichungen, 1970.* This publication is substantially a guide to the abbreviations used by organisations in Germany and abroad but the alphabetical sequence of abbreviations includes many standard prefixes including virtually all of the ones used by national bodies. In the absence of a completely satisfactory guide it will be necessary for the individual standards engineer or technical librarian to compile his own card index of standard and specification prefixes. To be effective this index must contain: 1 Prefix of standard or specification; 2 Issuing body; 3 Sources from which the standard may be obtained.

Prefixes contained in a similar index have been enumerated in *Proceedings of the engineering materials and design conference* held on 22-26 February 1960 at Earls Court, published by Heywood and Co. The index was originally compiled by Mr R K Whitehead, standards engineer at English Electric Co, and it could well be used as the basis for any personal index.

The Ministry of Defence (First Avenue House, High Holborn,

London, WC1) has compiled and issued in mimeographed format a useful guide to prefixes used in government departmental standards excluding specifications for major equipment such as vehicles and guns. The list is in tabulated form and gives:

Reference letter	Numbered series	Subject of specifications	Technical, approving or issuing body
AERE	I-onwards	Requirements for atomic energy use	Atomic Energy Research Establishment, Harwell, Berks

A well known central index of standards and specifications devoted to a single subject is the SINTO *Index of steel specifications,* which is housed at the Sheffield Commercial and Technical Library and now contains over 15,000 entries.

STEEL SPECIFICATIONS

Steel and metallurgical specifications are an exception to the general rule in that they are usually well documented. A useful guidebook to steel castings specifications is issued by the British Steel Castings Research Association under the title *British and foreign specifications for steel castings* (third edition, two volumes 1968). This lists foreign standard specifications and their British equivalents. The Comité de Liaison des Industries Metalliques Européennes (COLIME) has issued *Stahlvergleich der Gangigen Sorten* with its text in five languages. This is arranged by type of steel, then by country, and enables each steel purchaser to select the steel characteristics which will most nearly satisfy his needs in the various steel producing countries of Europe. It is also intended to serve as a guide until unification of steel specifications is achieved through ISO. Both the above publications are invaluable for helping the enquirer who is seeking the British equivalent of a foreign steel specification. *Engineering alloys* by N E Woldman (Reinhold, fourth edition 1962) lists some 35,000 alloys manufactured in the United States and abroad, giving details of their chemical composition and physical properties. The main sequence giving data on and uses of each alloy is arranged by a key number and is supplemented by indexes of alloy trade name (which corresponds to a specification prefix) and manufacturers' names. *Stahlschussel: key to steel* (Marbach A/N, Verlag Stahlschlussel) gives information on the composition, pro-

89

perties, uses, etc of various types of steel manufactured in Germany and throughout the world. *Steel's specification handbook* is a cross-index of chemically equivalent specifications which groups together specifications for materials of similar chemical composition; these are arranged under a series of code numbers which are assigned to each group of specifications. A second sequence lists the specifications by their issuing bodies, *eg* ASTM, SAE, then by specification number.

US STANDARDS

The identification of standards and specifications is made easier by two important publications of NBS. The *National directory of commodity specifications* (MP 130, 1945) is a compilation of 1,300 pages which is useful for tracing older specifications; its subtitle 'classified and alphabetical list and brief description of standards of national importance' is an adequate description of its scope and arrangement. More recently, NBS have published *Directory of United States standardization activities* compiled by Joan E Hartman (MP 288, 1967). This inventory of the work of about 486 American organisations involved in standardisation is divided into government and non-government agencies.

E J Struglia's *Standards and specifications : information sources* (Gale Research Co, 1965) lists directories and indexes of standards issued in the United States and gives information on the standards available from governmental and independent industrial sources.

FILING

Standards are invariably filed first of all by country then by issuing body and standard number. As standards and specifications are usually requested by their number, it is not common practice to catalogue or classify these documents; the filing order on the shelf should reveal whether the library holds a particular standard. The subject approach to standards can be met by reference to the yearbooks issued by national and international bodies. Some librarians will use marked copies of these publications to obviate the need for cataloguing and to indicate which standards the library holds without reference to the shelf order.

Chapter nine: Technical report literature

A technical report is an account of work done on a research project, which a scientist compiles to convey information to his employers and also to other scientists working in the same or a related field. Reports are also produced to record the results of research for patent purposes and for future use. The document may be either a final report on a completed investigation, or one of a series of reports which are issued at intervals to show how work is progressing on a particular investigation.

Scientists have always acknowledged the need to write up their experiments in order to communicate the developing state of their art to other scientists. Rutherford did not consider a discovery complete until it had been described in simple and direct English, and deplored the attitude of those who spent months of intensive research on a subject but begrudged a few hours work on the presentation of the facts for publication. It has been suggested that every US government sponsored research programme should have a definite clause in its contract detailing reporting requirements, to avoid the loss of valuable information through non-reporting.[1]

Technical report literature is an evolution of older methods of disseminating and recording the results of scientific research— personal communication from one scientist to another, publication in the journals of a learned society and the writing up of laboratory notebooks. Personal communication was, perhaps, adequate in a more leisurely age when one scientist was able to know exactly who was working within his field and was able to inform them of his researches simply by writing a letter. The present magnitude of scientific and technical research now makes such methods of communication impossible. It has been estimated that some ninety percent of all the scientists who ever lived are living now.

Present day research is undertaken not by individuals but by teams of scientists working together in the laboratory of a government research station, a university or an industrial organisation.

The projects under research are usually individual segments of a much larger programme within the organisation's range of activity. The technical report is a more efficient method of recording the results of an investigation than the laboratory notebook, because it serves to disseminate these results to anybody who receives a copy of the report.

HISTORY OF THE REPORT

One of the earliest series of technical reports to be issued in Great Britain was the ' R & M (research and memoranda) series ' of the Advisory Council for Aeronautics, which commenced in 1908. In the USA the various series of the US National Committee for Aeronautics were issued from 1918. Report literature has developed as a vehicle of communication between scientists, and become established as an alternative method of publication to the scientific periodical, only since the start of the second world war, when the pattern of scientific and technical research was profoundly altered by government intervention for reasons of national defence. During the war the US and other governments poured millions of dollars into the development of new industries. The results of this research could not be published in open technical literature for security reasons, and thus the technical report became an accepted medium of communication. Whole industries, such as aeronautics and atomic energy, were built almost exclusively on the information contained in reports. Government interest in scientific and technological research did not cease in 1945 but expanded beyond purely military fields. The second industrial revolution of automation science-based industries—chemical, electrical, aircraft—has involved governments of all countries in industrial science. Science is no longer an academic pursuit, but a matter of national survival.

Research by industrial firms has increased accordingly. Whereas in 1900 it was extremely rare for a scientist to be engaged on research outside a university, today no large industrial organisation is without its research department. In the United Kingdom alone national research expenditure has increased from twenty or thirty million pounds in the late 1940s to £962.1 million in 1967/8.

The phenomenal increase in scientific and technical research produced a parallel increase in scientific publications. The traditional channel of communication, the scientific periodical, could no longer keep pace with the sudden 'information explosion' and much information was of necessity released in technical reports. The report, which is usually produced by one of the near-print processes—offset-litho, gestetner etc—has one great advantage over the scientific periodical as a medium of communication, in that it can make the information available with a minimum of delay. Scientific and technical periodicals quite commonly have a backlog of papers awaiting publication for twelve to eighteen months.

An important body of reports which were issued after the second world war were the BIOS, CIOS, JIOS and FIAT series, containing the results of German and Japanese wartime research as edited by allied research teams from captured enemy documents. The demand for these reports did much to make librarians aware of the existence and importance of report literature. The *Weinberg report*[2] estimated that 100,000 government reports are issued each year in the USA alone. Most of these documents are issued by government research departments or by industrial organisations under government contract.

EDITING REPORTS
Some indication of the current importance of the reports can be measured from the publication of numerous manuals on technical report writing and editing. The librarian or information officer will often be responsible for editing the technical reports issued inside his organisation. He will co-operate with the scientific author to ensure that the report conforms to a set 'house style', that the information has been presented clearly and logically and that any bibliographical references have been correctly cited. Of forty one information/library units replying to the ASLIB survey,[3] fourteen indicated that they undertook some editorial, bibliographical or distribution work on their companies technical reports. The *Weinberg report*[2] underlined the need for the editing of technical reports, and indicated that although technical reports were not originally conceived as of archival permanence, the growth in government report publishing has tended to formalise the technical report giving it this status. It recommended that editors who are technically competent and in touch

93

with authors' work should be appointed by report issuing bodies to ensure that only material of worth is documented.

CLASSIFIED REPORTS

The technical report became accepted originally because it was virtually the only means of documenting much security information. This information is contained in ' classified ' reports. Security in the context of report literature has been defined as the ' safeguarding and protection of classified documents against unlawful dissemination, duplication, or observation because of their importance to national defence '.[4] ' Classified ' refers to the degree of secrecy which prevents disclosure to unauthorised persons. Each document is security classified individually, ' subject to future change or declassification '.[4] It has been noted that greater effort should be devoted to eliminating unnecessary over-classification of report literature at the time of generation, as this would obviate the time consuming task of downgrading or reclassification after publication.[1]

The foregoing definition of a classified report applies to government sponsored documents. Many internal reports, however, produced in the research department of an industrial firm will also be classified by the firm; these reports will contain vital information concerning company projects and processes which are to be available only to senior scientific and technical company staff. The degree of classification of an internal company report will not normally be decided by the author. Jermy has pointed out that the technical expert can not be expected to be familiar with all the commercial implications of his work.[5] Classification is invariably undertaken by a committee of experts representative of senior management and research staff whose experience qualifies them to assess both the technical and commercial implications of a research project.

TECHNICAL REPORTS AND PERIODICAL LITERATURE

Several surveys have been undertaken on the fate of the information contained in technical reports. A survey carried out jointly by the National Science Foundation (NSF) and the Library of Congress (LC) in 1952 indicated that most of the information contained in unclassified government reports eventually appears in scientific and technical periodicals. Of eighty four reports

94

investigated, only one that was not to be published contained information suitable for publication; the material in eighteen reports was unsuitable for publication; the information contained in the remaining sixty five reports had either been published or arrangements had been made for its publication. This survey did not contain any details of the delay between the writing of a report and publication of the information it contained in open literature, but in view of the delay previously noted in journal publication, it is evident that scientists and technologists will continue to be dependent on report literature for the rapid dissemination of much technical information.

A more searching survey of 2,500 reports issued during January to June 1952, made by the Technical Information Division of LC, indicated that about sixty five percent of unclassified technical reports contain publishable information; that all the information in about fifty percent of these reports is published in the open literature within two to three years; and that in the case of about twenty percent of the reports which contain publishable data, the information is not published for several years, much of it never appearing in a conventionally printed form. This survey also indicated that one report only seldom leads to exactly one published paper. More often a single journal article will include important results from several related reports or a single report may provide data for several published papers.[6]

The US Air Force Office of Scientific Research and the National Aeronautics and Space Administration both estimate that between fifty and sixty percent of their unclassified reports are eventually published in the open literature.[7] Some US government agencies which sponsor unclassified research give the investigator the option of publishing his results in the appropriate journal, and will not issue these results in report form unless the investigator authorises this form of publication, since it is believed that journal publication is a more effective means of disseminating information than technical report literature.

Reasons for the non-publication of the information contained in technical report literature are:

1 The bulk of some reports and the high cost of journal publication (although some US government agencies will pay the page charges to research journals for the publication of important

papers, technical reports, or papers based on them which have been the result of government sponsored research).

2 The highly specialised nature of much research which would have a limited interest.

3 Many technical reports are unedited and would need ' cleaning up ' for publication, involving time which the author could not spare.

4 The confidential nature of much of the information contained in report literature.

5 The data obtained as a result of the research project may be insufficient to make a paper worthwhile.

REFERENCES
[1] Kee, W A: *Sci-tech news* Spring 1962 7-9.
[2] *Science, government and information; the responsibilities of the technical community and the government in the transfer of information* (US Government Printing Office, 1963) 39-40.
[3] *Survey of information/library units in industrial and commercial organizations* (ASLIB, 1960) 41.
[4] Fry, B M: *Library organization and management of technical reports literature* (Catholic University of America Press, 1953) 10.
[5] Jermy, K E: *Aslib proceedings* 18(8) August 1966 218-223.
[6] Gray, D E and Rosenborg, S: *Physics today* 10 June 1957 18-21.
[7] Herner, M and Herner, S: UNESCO *Bulletin for libraries* 13 (8/9) August-September 1959 191.

Chapter ten: Problems of handling technical report literature

*T*he chief problems encountered in the handling of technical report literature are: 1 Lack of bibliographical control; 2 Identification; 3 Availability; 4 Cataloguing; 5 Storage and filing; 6 Security matters.

BIBLIOGRAPHICAL CONTROL

Technical reports are not listed in the book trade bibliographies such as *CBI* and *Whitaker,* or in specialised book selection aids such as *Aslib Book list* or *New technical books,* although some government reports issued in the UK by the Department of Trade and Industry's research stations and the United Kingdom Atomic Energy Authority, which are on sale through HMSO, are included in the HMSO lists and appear in the *British national bibliography.*

Some US government reports which are not listed in US *Government reports announcements* also appear in the *Monthly catalog of US government publications.*

Some technical periodicals also include lists of reports which would be of interest to their readers, examples being such periodicals as *Aircraft engineering* and *Engineering.*

The major abstract journals have tended to ignore reports which lie outside ' well defined and easily identified series '. The following reasons have been suggested for the non-coverage of reports by conventional bibliographical tools: [1]

1 Existing services are barely able to keep up with conventionally published literature, and as report literature is so difficult to obtain it would be impossible to control efficiently.

2 Even if report literature were covered by abstract journals it would be extremely difficult for the reader to obtain many reports listed.

3 Reports have a relatively short life and in many cases are not worth listing; some report libraries automatically discard reports after four years.

4 Unedited technical reports in which the material is not adequately presented should not be considered as part of the open literature. A survey of the coverage of technical reports by several abstract journals showed very sporadic treatment of this literature.[1]

REPORT ABSTRACT JOURNALS AND BIBLIOGRAPHIES

Several specialised abstract journals, which are mainly of US origin, have been established in an effort to bring report literature under bibliographical control. US *Government reports announcements (GRA)* is issued twice a month by the National Technical Information Service (NTIS) in order to announce the availability of new reports of government sponsored research and development released by the Department of Defense (DOD), the Atomic Energy Commission (AEC), the National Aeronautics and Space Administration (NASA), and other agencies. GRA is arranged under twenty two subject categories each of which is again more specifically subdivided. Within these groups the reports and translations are listed with full bibliographical citations, indicative abstracts and a string of indexing terms which can be adapted for use in retrieval systems. Each issue contains a report number index. The *Government reports index* (GRI) is published twice a month by NTIS and is the key to the report literature abstracted in GRA. GRI provides access to the reports under corporate author, personal author, subject, contract number and accession/report numbers. NTIS has also made available over 200 *Selective bibliographies* and *Catalogs of technical reports,* covering many areas of broad interest such as information retrieval, transistors, lasers, etc. A list of these report bibliographies is available on request from NTIS at the US Department of Commerce, Springfield, Virginia 22151. Additional NTIS facilities include the *Fast announcement service* (FAS), covering fifty seven subject categories, which processes approximately ten percent of the NTIS document input to give a rapid service emphasising the commercial applications of government sponsored research and development and *Government reports topical announcements* (GRTA), a semi-monthly current awareness service published in thirty five separate fields of technology and designed to aid the busy scientist, executive or engineer by summarising new technical information in a convenient form for easy assimilation.

Scientific and technical aerospace reports (STAR) is issued twice a month by the scientific and technical information division of NASA, and is a comprehensive abstracting and indexing journal covering world wide report literature on the technology of space and aeronautics. Each issue contains subject, corporate author, personal author, report/accession number indexes. Separate cumulative indexes are published quarterly, with the exception of the last quarter of the year, which is an annual index.

Nuclear science abstracts (NSA) is issued twice a month by the United States Atomic Energy Commission to announce unclassified reports published by the commission. NSA also covers world wide periodical literature and unclassified nuclear science reports issued by other national government agencies. Each issue contains corporate, personal, subject and report number indexes which cumulate quarterly, semi-annually, annually and quinquenially. Unclassified reports issued by the United Kingdom Atomic Energy Authority (UKAEA) are listed by prefix in the monthly *List of publications available to the public.*

The International Nuclear Information System (INIS) which is sponsored by the International Atomic Energy Agency in Vienna, made its *INIS atomindex* service available in May 1970. The service offers a computer printout, grouped according to a subject classification, giving complete bibliographical descriptions supplemented by keywords of all items acquired by the INIS library.

The National Lending Library for Science and Technology (NLL) began issuing a monthly classified listing entitled *British research and development reports* in January 1966. This has now been absorbed in the *NLL announcements bulletin* which includes brief bibliographical information about the reports acquired by the library.

R and D abstracts is published twice a month by the Department of Trade and Industry's Technology Reports Centre (TRC) and includes abstracts of British and foreign reports inside twenty two subject groupings. Each issue contains an average of 300 abstracts and any item included may be obtained on loan from TRC. From 1970 abstracts covering certain subjects have been issued separately to form four page bulletins catering for particular interests such as aerospace engineering, materials and manufacturing methods, and physics.

Although most of the information contained in the BIOS, FIAT,

CIOS and JIOS reports has now been published in the open literature or has been superseded by technical progress, these documents are still occasionally requested and their indexes, *Technical index of reports on German industry* in six parts plus a supplement, and *Reports on German and Japanese industry* in twenty classified lists, both of which were originally issued by the Technical Information and Document Unit of the Board of Trade can be consulted in most large public technical libraries in the UK.

Lists of reports are issued in the UK by most of the research stations and research associations and are usually available on request or exchange. Examples of these lists are the National Engineering Laboratory monthly *Lists of reports* and the British Iron and Steel Industry Research Association bi-annual *Open report list,* which contains abstracts of reports available for general circulation which have originally appeared in the *Restricted report list,* issued six times a year and available to members only. A useful publication which gives information on the research activities of all research associations and stations, and which includes details of reports lists and series issued by these organisations, is *Technical services for industry,* published by the Department of Trade and Industry in 1970. *Industrial research in Britain* (sixth edition 1968) edited by I D L Ball is another valuable source of information on the report literature issued by independent and sponsored research laboratories, trade and development associations and professional and learned societies.

The ' Science Information Exchange ' is administered in the United States by the Smithsonian Institution, and exists to inform individual investigators about who is working on problems in their special fields. Information is received by the Exchange from both government and non-government sources in the USA and a 200-word technical summary of each project is produced. These ' notifications of research projects ' (NRPs) are then indexed, processed and stored in a computer for subsequent retrieval; details of over 100,000 projects are received each year. By using this service to locate research activity in a specific field, an information worker can acquire reports relating to the research, if these are freely available, by personal contact with the organisation conducting the work. The establishment of a similar service in the UK, perhaps under the auspices of OSTI, would be of great value to industry.

IDENTIFICATION

Reports are usually referred to by readers and cited in open literature by report code numbers. It is less usual for a reader to ask for a report by the author or title, although an enquirer may wish to know which reports have been issued on a specific subject by his research department. The report number is assigned to the document by the originating organisation, to act as a brief means of identification, giving a shorthand description of the origin and status of the document. The best-known report designations are probably the AD (ASTIA document), PB (Publications Board) and TID (Technical Information Division) prefixes assigned to their publications by the US Armed Services Technical Information Agency (now the Defense Documentation Centre), the US National Technical Information Service and the US Atomic Energy Commission. The report number can also facilitate filing and can even aid security by concealing the name of the author when the document is cited. The report number may consist of, or be a combination of any of the following units:

1 The initial letters of the name of the issuing organisation, eg MIT—Massachusetts Institute of Technology; AERE—Atomic Energy Research Establishment.

2 An indication of the form of the document: TN—technical note; PR—progress report.

3 An indication of the status of the document, eg C—classified, S—secret.

4 The date of writing the document, eg 01-1-57—1 January, 1957.

5 The name of the project which the document reports.

6 An indication of the subject content of the document, eg H—health; LS—literature search.

Within any individual series of reports the documents are usually assigned a running number; thus NACA-WR-A-6 may be decoded as the sixth *Wartime report* issued by the Ames Laboratory of the National Advisory Council for Aeronautics.

Identification of a report means ascertaining from the report number which organisation issued the document. This is a difficult (often impossible) task because there is no adequate control over the assignment of report codes. Reports issued by the same authority may change their prefixes if the organisation changes its name, or the same prefix may be used by two entirely unconnected organisations. Again a reader who is ignorant of the problems of identification may add to the confusion by abbreviating a number, or by omitting any of the component parts of the number when requesting the document.

Although numbers are usually assigned by the issuing body, this is not always the case. In the absence of a number, the librarian of a company which maintains a reports collection might prefabricate his own number for a series of reports for filing purposes, by assigning a number consisting of the initial letters of the organisation and a running number. The librarian may have unwittingly increased confusion by producing this hybrid report number, which may thereafter be cited in the open literature by anyone who has consulted the report in his library, yet be completely unintelligible to anyone outside the organisation. Difficulties are increased by the indiscriminate use of such codings as TR—technical report or PR—progress report.

A useful tool which helps to overcome the identification problem was published by the Special Libraries Association (SLA) in 1962. *The dictionary of report series codes* lists 12,500 codes used by 4,000 organisations. The various report series references used by the UKAEA are listed in the appendices to J Roland Smith's *Guide to UKAEA documents* (fourth edition, UKAEA, 1966). The number of report codes is growing so rapidly, however, that it will be necessary for any librarian handling reports in any quantity to maintain his own card index to supplement the dictionary.

A proposed solution to the identification problem in the United States has been made by Helen Redman, librarian at the Los Alamos Scientific Laboratory, New Mexico, who suggests that SLA should adopt a statement covering the following points:

' 1 The requirement that every report, whether formal or informal, final or interim, distributed or strictly internal bear a report number.

' 2 The necessity for that number to be a unique identification of a report, and the desirability of its being brief and simple.

' 3 A listing of essential components of report designation, such as a code for the originating, sponsoring, or distributing agency plus an identifier of the specific report.

' 4 A recommendation of format, including the order of components and the spacing and punctuation between them.

' 5 A permissive statement about other components that might be added to the essential designation '. Mrs Redman then suggests that a central authority file of codes which are already in use should be maintained.[2]

AVAILABILITY

The majority of technical reports are not available through normal book trade channels and can only be obtained through a government clearinghouse or lending service, directly from the originating organisation or even by contacting the author of the report.

In Great Britain most government sponsored unclassified reports can usually be purchased from HMSO. ASLIB will endeavour to locate and obtain on loan any specific government or non-government report which is freely available, on behalf of its members. NLL, Boston Spa, Yorkshire maintains a loan collection of: reports which include AD and PB reports issued by NTIS covering work carried out under US government contract; atomic energy reports from over forty countries and from international organisations such as the International Atomic Energy Agency and European Nuclear Energy Agency; all reports issued by NASA; RAND reports issued by the Rand Corporation of Santa Monica, which cover a wide variety of subject fields; the second world war FIAT, BIOS, CIOS and JIOS reports; and a broad range of British and foreign research and development reports, issued by universities and industrial organisations. It will be appreciated that NLL's collection only contains unclassified reports. The Department of Trade and Industry's Technology Reports Centre (TRC) at St Mary Cray in Kent acts as a clearinghouse for both the 8,000 reports generated each year by the department's research and development establishments and also for those acquired from overseas sources. Overseas reports, accessioned at the rate of 40,000 per year, form the major part of the collection, with the USA being the main supplier. TRS holds impressive collections of US reports from 1941 onwards including many of the war-time reports issued by the US Office of Scientific Research and Development. Its collection of AD reports, particularly those numbered up to 255,000 is the most comprehensive in the United Kingdom. TRC holdings in part complement those of the NLL in that many of the classified reports not available from Boston Spa can be obtained from St Mary Cray by those who can demonstrate a 'need to know'. The centre is also the UK distribution point for reports issued by the NATO Advisory Group for Aerospace and Development (AGARD). Another category of publication held by TRC is the US *Military specifications* (see page 72) which are received at the rate of 6,000 per year.

In the United States the Department of Commerce's National Technical Information Service (NTIS) exists to coordinate the department's business and technical information activities and to serve as the primary focal point within the federal government for the collection, announcement and dissemination of technical reports and data. The reports processed and distributed by NTIS are the result of research work carried out by over 150 agencies within the federal government and their subject scope embraces all areas of scientific, social and economic research. NTIS also acts as a clearinghouse for information on current government sponsored unclassified work. Through its offices the public can locate relevant report literature on a particular topic or enlist the subject experience of specific centres of expertise.

The librarian or information officer must discover which organisations are pursuing research related to his own organisation's field of activity and, if this information is documented in technical reports, find out whether his organisation is eligible to receive these reports.

Government sponsored reports are usually distributed on an automatic basis and some will be available on request. The originators of a government sponsored research report may be furnished with an approval distribution list of individuals and organisations who are eligible to receive the report by the government department issuing the contract. The USAEC has a breakdown of twenty six subject categories for its technical reports, each with its own automatic distribution list. In the United States importance is attached for security reasons to the 'need to know', the principle being that an individual should only have access to as much information as is necessary to perform his particular job.

As only a limited number of copies of each technical report is duplicated, and this small number may soon become exhausted, it is often necessary to approach the author direct for a copy of his report if this is not available from the issuing organisation.

The results of government sponsored research which is unclassified from a security point of view may nevertheless be unobtainable, because although the government has the right to use the results of the research, the information may contain patentable material which will be the property of the organisation carrying out the research.

Reports produced by private industrial firms are not generally available and are restricted to the company's confidential file. The information contained in company confidential reports usually results in a patent application. It has been estimated that in certain competitive fields, *eg* detergents and powder metallurgy, only five percent of research information is made available to the public by industrial firms, and the bulk of this is contained in the patent literature.[3]

CATALOGUING

Reports present difficulties for the cataloguer, because so many separate items of information may appear on the title page: report number, issuing body; personal author; contracting agency (if any); security classification—each laying claim to the main entry heading. Organisations holding large collections of reports will usually select the name of the issuing body to take the main entry because:

1 The corporate name is more significant than the personal author in that it is more often used as a method of referring to the document (prefixes are usually indicative of issuing bodies).

2 A report is very often the work of several scientists and the author cited first on the title page is not necessarily more important than those following.

3 When an author leaves an organisation his work on a project will be continued by another scientist, and thus a final report and a progress report on the same project might appear under the names of different authors, which would be separated in the catalogue if personal names were adopted as main entry headings.

If entry under corporate author is adopted, the problem of whether to enter under the main organisation or under the department or laboratory name will have to be resolved, *ie* whether to make the main entry under ' Great Britain, Ministry of Technology, Radio Research Station ', or under ' Radio Research Station, Great Britain '; under ' Pittsburgh University, Mellon Institute ', or under ' Mellon Institute '. Detailed guidance on this problem is given in the Committee on Scientific and Technical Information's (COSATI) *Standard for descriptive cataloging of government scientific and technical reports,* AD 641092. The standard recommends that even if three or four organisational elements of a corporate body are given in a document only two elements should be chosen for the corporate heading, the largest followed by the smallest element, *eg* ' General Motors Co, Cincinnati, Ohio. Nuclear Materials and Propulsion Operation '. Adherence to the COSATI standard will ensure consistency of practice in a reports index.

An added difficulty in cataloguing reports will be the choice of title. Many reports are issued with several title pages and various forms of title which may include an overall project title and a specific document title. The titles of some reports are long and involved, and reports issued in the form of memoranda may have no titles. In such cases the cataloguer must choose the title which is most descriptive of the subject matter of the document, abbreviate the title, or manufacture a suitably descriptive title. Titles which commence with such phrases as ' final report ', ' progress report ' are often inverted to bring the significant words to the fore; thus, *Metal fatigue in aluminium track rods, final report on.* In these and many other cases the reports cataloguer must exercise his ingenuity in departing from conventional cataloguing practice in the interests of the user.

The COSATI standard cited above is the most comprehensive guide to the cataloguing of reports. It was formulated to ' 1) provide rules for descriptive cataloging appropriate to the needs of information centers, documentation centers, and reports departments of libraries; 2) provide users with a consistent form of citation and index entries among the products of the various information systems; 3) enable government agencies to use each others descriptive cataloging entries with minimal editorial review, thereby avoiding costly duplicative effort; 4) provide a guide to other organisations and institutions who wish to formulate compatible descriptive cataloging practices '. The rules cover the treatment of the various report elements including accession numbers, corporate authors, title, descriptive note, personal authors, date, pagination, contract number, report number, availability, supplementary note and security classification.

Elaborate report catalogues will only be maintained in large organisations whose interests are primarily covered by report literature. Most other organisations usually rely on report number filing order to locate reports issued by an individual organisation. If accession number filing is used, a report number conversion index which links the report number and the accession number will be needed. All libraries will need subject indexes to exploit the information contained in their reports collection, as the work of any organisation can only proceed in the light of research which has already been completed and is documented in the reports files. Post co-ordinate indexing is now widely used for the subject organisation of the information contained in report literature. Public libraries which maintain large collections of reports on atomic energy will not compile subject indexes to this material, but will rely on published tools such as *Nuclear science abstracts* to exploit their collections.

STORAGE AND FILING

The volume of reports now emanating from a wide variety of sources causes storage difficulties for any library maintaining a report collection. The space problem must be alleviated by ' weeding ', which is essential if the collection is not to contain obsolete and duplicated material. As previously noted, much information contained in technical report literature appears in open literature within two to three years. Progress reports will be superseded when the final report on the project is issued. The obsolescence of report literature is discussed in a letter by C W J Wilson published in *Aslib proceedings*.[4] The attitudes of librarians in the US aerospace industry to the maintenance of a report collection have been summarised by J H Richter.[5]

As no physical standards exist for technical reports, these documents appear in a variety of shapes and sizes, which aggravates the storage problem. It should be noted that sheet microfilm or microfiche is an economic method of publishing those technical reports which have a limited appeal. A project sponsored by the National Science Foundation has indicated that microforms are a feasible method of primary publication for information of interest to a limited clientele.[6] The USAEC is now issuing all its technical reports on microfiche; in addition to helping to solve the storage problem for report literature, this method of publication enables an enlarged hard copy to be produced from the fiche.

Several methods of housing reports are in use. Libraries with only modest collections of reports will usually adapt standard library shelving by dividing each shelf at intervals of perhaps six inches with vertical partitions to form pigeon holes, which will ensure that the reports do not ' spill ' over the shelf. In other libraries reports are filed between stout boards which are cut flush to the size of the reports and punched with holes through which tapes can be threaded. Each set of boards will secure a dozen or so reports into a homemade bound volume. A more suitable method of filing reports, which overcomes the difficulty of untying the tapes to gain access to a particular document, is to use mill boards cut to size and secured by a stout strap to enclose a number of reports. This method has been used successfully in the Department of Trade and Industry Library. Transfer cases are also commonly used for shelving reports and are more adaptable than sets of boards, as the individual boxes can be labelled on their spines with the report prefixes. The disadvantage of these methods of filing is that they will not adequately accommodate reports of varying size. Libraries with extensive collections will use either vertical filing cabinets or lateral wall filing units. The latter method of filing is to be preferred, as it has the advantages of immediate access (no filing cabinet doors to be opened or closed) Each pocket can be individually labelled, giving immediate identification of each report, access is not limited to the contents of one drawer at a time and lateral filing units are more economical of space than vertical cabinets. Any library in which wall space is at a premium but where floor space is available may use vertical filing cabinets, although it should be noted that ' island ' lateral filing units are available. If vertical filing cabinets are used, care should be taken that floors are not overloaded.

ARRANGEMENT

There are three obvious methods of arranging a collection of technical reports: 1 By report number; 2 By accession number; 3 In a classified sequence.

Most collections of reports are arranged by report number because:

1 As reports are usually cited by their number, such an arrangement gives the librarian or user direct access to the documents—

he has no need to use the catalogue or to compile a conversion index listing the report numbers against accession numbers.

2 As most report prefixes are indicative of the origin of the document, filing by report prefix will collocate reports issued from a single source.

3 A frequent approach to report literature is by source. For example, a scientist may start a search by citing several laboratories which are working on his subject and ask the librarian to produce reports from these organisations. In this case, filing by report number will facilitate retrieval of the documents. One disadvantage of filing by report number is that it is wasteful of space, since room must be left for expansion between each sequence of reports.

Arrangement by accession number is common practice because:

1 It is economical of storage space, since reports can be filed ' tight '.

2 If a system of post co-ordinate indexing, which uses a running number to identify each report, is used to exploit the collection, the librarian will have direct access from the index to the reports.

3 It facilitates the weeding of the collection in that older reports can be easily identified.

Disadvantages of filing by accession number are that this order denies access from the most common approach for external reports, that of report number, and that this order introduces an additional number by which the report may be cited. If accession number filing is adopted, a conversion index from report number to accession number will be needed.

Due to the highly specific nature of the contents of most technical reports, and the rate of development of many technical fields, classification is not widely adopted because of the difficulties involved in compiling and maintaining adequate classification schedules.

SECURITY
Most industrial libraries maintain a file of their own internal company reports, which will be housed separately from external reports. These documents are not usually available outside the organisation, and indeed reports of vital importance to the company will only be available to senior technical personnel within

the company. The internal reports collection is really the company's corporate memory which records the history of its technological development. J C Hartas maintains that a ' lack of internal documents is the most serious fate that can befall an information service, because it cuts out the most important information which no amount of published material can replace '.[7]

The security problem is approached in one company with a collection of several hundred thousand internal reports, which are kept on closed access, by classifying each report into one of three security grades: A Top security; B Documents containing information of vital importance to the company; c Other reports. Reports graded c and B can only be issued to readers who can produce the requisite report-issuing chit. These chits are issued to individual scientists in book form only on the authority of a technical director; thus a scientist who holds only a book of B chits cannot have access to a report with an A security classification. This procedure releases the librarian from the difficult task of having to refuse a scientist access to a classified report and the system is understood and accepted by all company employees.

As few readers would wish to browse among a report collection and open access is not essential, the system described integrates three grades of report into one sequence arranged by accession number. It is effective, for it guarantees that a report can only be filed in one place, no matter what its security classification may be.

Other approaches to the handling of confidential reports were summarised in the August 1966 issue of *Aslib proceedings*.[8, 9]

Organisations with relatively few classified internal reports will simply segregate these documents from unclassified reports by filing in locked cabinets, while keeping other reports on open access.

In organisations where vital security work is undertaken, *eg* atomic energy establishments, the most elaborate checking procedures will be made on highly sensitive documents which will usually be housed in a separate building or room-safe. These documents will be monitored in all of their movements inside and outside of the establishment: the security arrangements will usually include logging systems and periodical auditings of the individual copies of a single report.

REFERENCES

[1] Herner, M and Herner, S: *Unesco bulletin for libraries* 13 (8/9) August-September 1959 192.

[2] Redman, H: *Special libraries* 53 (10) December 1962 574-8.

[3] Fagerhaugh, K H: *American documentation* 3 1952 144.

[4] Wilson, C W J: *Aslib proceedings* 16 (6) January 1964 200-1.

[5] Richter, J H: *Sci-tech news* Fall 1967 64-5.

[6] Herman, C M and Davis, D E: *Bio-science* 14 (4) April 1964 27-30.

[7] Hartas, J C: *In* Houghton, B (ed): *Information work today*, 79-87, Bingley, 1967.

[8] Gerrard, S A and Lyle D F: *Aslib proceedings* 18 (8) August 1966 206-17.

[9] Jermy, K E: *Aslib proceedings* 18 (8) August 1966 218-225.

INDEX

113